Sustainability in Project Management

Reviews for *Sustainability in Project Management*

'Sustainability in Project Management thinking and techniques is still in its relatively early days. By the end of this decade it will probably be universal, ubiquitous, fully integrated and expected. This book will be a most valuable guide on this journey for all those interested in the future of projects and how they are managed in a world in peril.'
Tom Taylor dashdot and vice-President of APM

'Project Managers are faced with lots of intersections. The intersection of projects and risk, projects and people, projects and constraints... Sustainability in Project Management *is a compelling, in-depth treatment of a most important intersection: the intersection of project management and sustainability. With detailed background building to practical checklists and a call to action, this book is a must-read for anyone interested in truly implementing sustainability, project manager or not.'*
Rich Maltzman, PMP, Co-Founder, EarthPM, LLC, and co-author of *Green Project Management*, Cleland Literature Award Winner of 2011

'While sustainability and green business have been around a while, this book is truly a "call to action" to help the project manager, or for that matter, anyone, seize the day and understand sustainability from a project perspective. This book gives real and practical suggestions as to how to fill the sustainability/project gap within your organization. I particularly liked the relationship between sustainability and "professionalism and ethics", a connection that needs to be kept in the forefront.'
David Shirley, PMP, Co-Founder, EarthPM, LLC, and co-author of *Green Project Management*, Cleland Literature Award Winner of 2011

'It is high time that quality corporate citizenship takes its place outside the corporate board room. This excellent work, which places the effort needed to secure sustainability for everything we do right where the rubber hits the road – our projects – has been long overdue. Thank you Gilbert, Jasper, Ron, Adri and Julia for doing just that! I salute you.'
Jaycee Krüger, Member of ISO/TC258 a technical committee for the creation of standards in Project, Program and Portfolio Management, and Chair of SABS/TC258, the South African mirror committee of ISO/TC258

'Sustainability is no passing fad. It is the moral obligation that we all face in ensuring the future of human generations to come. The need to show stewardship and act as sustainability change agents has never been greater. As project managers we are at the forefront of influencing the direction of our projects and our organisations. Sustainability in Project Management *offers illuminating insights into the concept of sustainability and its application to project management. It is a must read for any modern project manager.'*
Dr Neveen Moussa, Project Manager, Adjunct Professor of Project Management and past President of the Australian Institute of Project Management

Sustainability in Project Management

GILBERT SILVIUS
RON SCHIPPER
JULIA PLANKO
JASPER VAN DEN BRINK
and
ADRI KÖHLER

Routledge
Taylor & Francis Group

LONDON AND NEW YORK

First published 2012 by Gower Publishing

Published 2016 by Routledge
2 Park Square, Milton Park, Abingdon, Oxon OX14 4RN

711 Third Avenue, New York, NY 10017, USA

Routledge is an imprint of the Taylor & Francis Group, an informa business

British Library Cataloguing in Publication Data
Sustainability in project management. -- (Advances in
  project management)
  1. Project management--Environmental aspects.
  2. Sustainability.
  I. Series II. Silvius, Gilbert.
  658.4'083-dc23

Library of Congress Cataloging-in-Publication Data
Sustainability in project management / by Gilbert Silvius ... [et al.].
    p. cm.
  Includes bibliographical references and index.
  ISBN 978-1-4094-3169-5 (pbk.)
  1. Project management. I. Silvius, Gilbert.
  HD69.P75.S87 2011
  658.4'04--dc23

2011042701

ISBN 9781409431695 (pbk)

CONTENTS

LIST OF FIGURES

LIST OF TABLES

LIST OF ABBREVIATIONS

AIPM   Australian Institute of Project Management

APM   Association for Project Management

CPM   Critical Path Method

CSR   Corporate Social Responsibility

EFQM   European Foundation for Quality Management

GDP   Gross Domestic Product

GE   General Electric

GRI   Global Reporting Initiative

HR   Human Resources

IPMA   International Project Management Association

ISO   International Organization for Standardization

NGO   Nongovernmental Organization

PERT   Program Evaluation and Review Technique

PMCD Project Management Competence Development framework

PMI   Project Management Institute

PMO   Project Management Office (or Portfolio Management Office)

PPP   People, Planet, Profit

TBL   Triple Bottom Line

UN   United Nations

WBS   Work Breakdown Structure

FOREWORD
NELMARA ARBEX

There was a time when sustainability was a topic for discussion among 'alternative' thinkers or 'softies'. This has changed dramatically. The social and environmental challenges that we are facing currently in our everyday life, everywhere around the globe, are the evidence that sustainability is and will be the most constant topic for all of us in the next century: citizens, politicians, entrepreneurs, investors, business leaders, experts, auditors, children, teachers, journalists, everyone... but I believe that this is particularly true for project managers.

The sustainability challenges are of a very concrete nature: how to develop and implement projects that guarantee the protection of the planet's natural resources and increasingly create wealth for more people? If we are unable to answer such a question we probably will not be able to guarantee access to water, clear air, food, health and safe living conditions to the next generations – our children and grandchildren. So a lot of what our future will be is in the hands of project managers.

The scale of industrial production is the size of the globe: all ecosystems and communities everywhere have been used for or modified by industrial activities. And the plans of all countries and businesses are to continue growing, as the current concept of development is based on a continuously growing production model, limitless.

However, the planet is a limited system, its resources are finite, or they need time to be restored. This planet is not compatible with unlimited and quick destruction. Our governments and society are also not able to cope with the size of inequalities created by the current development model. This is the core of the 'sustainability business case'. We will have to find new ways to develop and implement projects, to produce innovations and to find scaling-up solutions that can reverse the destruction caused by the current development model and consumption fever. Despite all current efforts – both to reduce destruction and to create totally new solutions – we are still destroying most of our natural resources in a permanent way, and we are creating more violence and poverty, as one can learn from the United Nations Reports.

When the authors told me about the idea of preparing this book, I gladly offered my support.

I can't imagine more critical professionals than project managers, project sponsors and project leaders to help us face these challenges. By their nature, projects enable change. Now we are living in a time when projects' concepts themselves must change if the project management field is to achieve its fullest impact as a driver of progress and improvement.

Not much has been done in this field. If one starts thinking about the size of responsibility in the hands of those who are designing and implementing projects everywhere, all the time, it is very easy to understand that they are the ones who have to be prepared and motivated to embrace sustainability as the most fundamental part of their work. This book is a contribution in this direction.

The task is not small and not simple. It is hard to imagine that any problem can be tackled by one of few experts. In order to face it, project managers and project leaders need a new way to perceive the world around them; they will have to be creative, they will need to think in a complex interconnected frame, they will have to understand their responsibility; they will have to develop new professional ethics, and will have to collaborate as never before. These new skills and understanding also lead to the need for developing new tools to manage the projects. Consequently, the new generation of projects needs new performance indicators to measure the achievement of the project's sustainability goals.

This book is a step in the direction of building the most needed professional skills of our time: creativity, responsibility and collaboration. I hope that after reading this book the reader will feel enthusiastic about the huge contribution they can make to our sustainable future.

Dr Nelmara Arbex
Deputy Chief Executive at the Global Reporting Initiative
(For more information and to contact Dr Arbex: www.sustentabilidadecompimenta. com.br)

FOREWORD
MILES SHEPHERD

Our world, since the earliest recorded time, has been evolving and changing: historians and anthropologists tell us how mankind moved from an animistic existence through various intermediary stages of physical and intellectual development until we emerged as the dominant inhabitants of the modern world. Our racial history is one of constant change and many observers would say that the rate of change has accelerated in the past 60 years or so since the end of the Second World War. There are many examples of the rapidity of modern change such as the development of consumer electronics and conveniences such as washing machines, cars and communications equipment such as the telephone. So mankind enjoys a much 'easier' life thanks to developments of many kinds, from consumer goods to medicine, wonder drugs that prolong life, increased food production and transportation. Other observers will remind us we owe a great debt of gratitude to the development of project management techniques which have allowed these wide ranging changes to transform our physical world. The discipline that emerged from the urgent needs of a world in crisis has brought the stars closer to us and has made many contributions to our society. So we enjoy a more comfortable lifestyle than our ancestors, we live longer, are healthier and have more leisure time. Or at least some of us do.

Still other observers will note that project management, as well as being an enabler for the advancement of mankind, brings with it some other characteristics – such as the consumption of raw materials. For the wind turbine cannot generate electricity without the concrete to build the tower, the metals to construct the generator and all the other specialist materials to complete the structure. Not only are raw materials needed in the production phase of the project but the project outputs frequently need resources to perform their function. And often, too, these materials are finite in their availability so projects contribute to the loss of precious natural resources.

Society has begun to regard consumption of limited resources as a significant issue and sustainability has become an all pervasive concept. While the case for sustainability appears compelling, there are many aspects of the concept that need to be put into context. In particular, responsibilities are far from clear. Just whose job is it to make sure that the project is 'sustainable'; what do we mean by 'sustainable'? These are two of many important questions that this book addresses. General advice on the meaning of 'sustainability' is provided by specialists in the Green movement but, for the project professional, the need to strive for sustainability goes beyond simple resource utilization. Project professionals react to the need for sustainability partly by conservation of resources but also need to consider other aspects such as the ethics of project objectives, the broader analysis of business cases, technical aspects of task execution and perhaps development of alternative project management methods.

There is little practical guidance beyond the mantra of consume less, reuse where possible and recycle more. There is even less guidance available on what is meant by sustainable project management, the responsibilities of the project management professional and less still on the critical role of the Executive Sponsor or their responsibilities. So this book, which places sustainability in both the business and the project management context, goes a long way towards providing this missing guidance.

Some may feel that the Project Manager has limited scope for injecting sustainability into the project but the very least that can be expected of the project professional is an awareness of the underlying principle of sustainability. The portfolio and programme managers need to move beyond this basic understanding towards a pragmatic implementation of the principles and this book provides guidance at all levels. Furthermore, advice for the Project Management Office is given so the Executive Sponsor, so critical in the corporate development of sustainability, can have every confidence in the project, portfolio or programme management team to achieve strategic objectives.

Miles Shepherd
Vice President, Assocation for Project Management

ABOUT THE AUTHORS

A.J. GILBERT SILVIUS

Gilbert Silvius (1963) is professor at HU University of Applied Sciences Utrecht in the Netherlands. He is Program Director of the first Master of Project Management programme in the Netherlands. This innovative programme focuses on project management from an organizational change perspective. The Master of Project Management has a special focus on the integration of the concepts of sustainability in projects and project management. Also in research, Gilbert focuses on sustainability in projects and project management.

In addition to being an established academic, Gilbert is an experienced project manager with over 20 years of experience in various business and IT projects. As a principal consultant at Van Aetsveld, Project and Change Management, he advises numerous organizations on the development of their project managers and their project management capabilities.

Gilbert is a member of IPMA, PMI and the ISO/PC 236 that develops the ISO 21500 guideline on project management. In the Dutch IPMA chapter, Gilbert is board representative for higher education.

RON P.J. SCHIPPER

Ron Schipper (1971) is project manager and consultant at Van Aetsveld, a leading consulting firm in project and change management in the Netherlands. He has more than 15 years of experience as a project manager in realising (organizational) change in various organizations. As well as executing projects, he is interested in developing the profession of project management and transferring this knowledge to other people in the Netherlands and developing countries. As sustainability emerges as a theme for the world, his attention has focused on the implications for projects and project management. Schipper is also an external examiner in the Master of Informatics programme at HU University of Applied Sciences Utrecht in the Netherlands.

JULIA PLANKO

Julia Planko works as a researcher and lecturer at HU University of Applied Sciences Utrecht since 2008. She lectures in Economics and Marketing and supervises students undertaking sustainability-related research projects.

Before relocating to the Netherlands she worked as associate lecturer at the Berlin School of Economics. Planko studied International Business Management and International Economics. For her Master's thesis she was awarded a DAAD scholarship to do field research in Madagascar on the topic of sustainable development. She developed her knowledge on this topic during her internship at the German Ministry for Economic Cooperation and Development in the department for political planning and development policy. Based on these experiences and insights, Planko is convinced that businesses have to play a major role in sustainable development.

She combines her passion for sustainable development and international trade with her interest in strategy and marketing in her current research topic. Her research focuses on how companies integrate sustainability into their strategy and how they can thereby impact the behaviour of other members of the values chain.

JASPER VAN DEN BRINK

Jasper van den Brink (1970) studied Social Sciences at the University of Utrecht (The Netherlands) and did a second master's study at the faculty of Law in Leuven (Belgium). He worked for many years as a project manager and a programme manager. In 2008 he started working at the HU University of Applied Sciences Utrecht where he conducts research on the application of modern psychological insights in project management. He teaches and trains subjects related to the behavioural part of (project) management such as project management skills, (team) coaching, and leadership. He is the owner of MWP Coaching, a coaching company in Utrecht, The Netherlands.

ADRI KÖHLER

Adri Kohler (1951) is a senior consultant/project leader involved in initiatives in ICT and business operations. In terms of education he was closely involved as a project leader in the development of the Master of Informatics and the Master of Project Management, where he is now course director.

He is also active at the HU University of Applied Sciences Utrecht within the Extended Enterprise Studies Research Group. In this capacity he is involved in the 'Care at a Distance' project, where the introduction of (information) technology for processes and organization of care establishments is central.

CHAPTER 1

INTRODUCTION

GILBERT SILVIUS, RON SCHIPPER AND ADRI KÖHLER

1.1   WHY THIS BOOK?

'Panta rhei' was the immortal wisdom, spoken over 2,500 years ago by the Greek philosopher Herakleitos. 'Everything flows'; everything changes and nothing remains the same. More than ever, this is true for the competitive environment of many organizations. Whether resulting from technological progress, new regulations, globalizing economy or inventive competitors, new developments change the marketplace every day. Organizations are continuously reacting to these changes, or anticipating new ones, by introducing new products and services, improving business processes, changing resources, expanding their activities or discarding obsolete activities. Selecting the right changes and organizing and managing them in an effective and efficient way is, for many organizations, a critical success factor for business agility and continuous success. Many of these changes are managed as projects: unique efforts that require the mobilization of resources of different disciplines, capabilities and organizational units. Project management is developing into the key organizational skill in order to execute these changes in a controlled manner.

However, an increasingly dynamic environment is not the only development organizations face. In the last ten to 15 years, the concept of sustainability has also grown in recognition and importance. For example, the pressure on companies to broaden their reporting and accountability from economic performance for shareholders, to sustainability performance for all stakeholders has increased (Visser 2002). The 2009–2010 world crises may even imply that a strategy focused solely on shareholder value is no longer viable (Kennedy 2000). Also following the success of Al Gore's 'inconvenient truth', awareness seems to be growing that a change of mindset is needed, both in consumer behaviour as in corporate policies. How can we develop prosperity without compromising the life of future generations? Proactively or reactively, companies are looking for ways to integrate ideas of sustainability in their marketing, corporate communications, annual reports and in their actions (Hedstrom et al. 1998; Holliday 2001). Sustainability, in this context, can be defined as: 'Adopting business strategies and activities that meet the needs of the enterprise and its stakeholders today while protecting, sustaining and enhancing the human and natural resources that will be needed in the future' (International Institute for Sustainable Development and Deloitte & Touche 1992).

The concept of sustainability has more recently also been linked to project management (Gareis et al. 2009 and 2011; Silvius et al. 2009 and 2010). Association for Project Management (APM) (past) Chairman Tom Taylor recognizes that 'the planet earth is in a perilous position with a range of fundamental sustainability threats' and 'Project and Programme Managers are significantly placed to make contributions to Sustainable Management practices' (Association for Project Management 2006). And at the 22nd World Congress of the International Project Management Association (IPMA) in 2008, IPMA Vice-President Mary McKinlay stated in the opening keynote speech that 'the further development of the project management profession requires project managers to take responsibility for sustainability' (McKinlay 2008). Her statement summarized the development of project management as a profession as she foresees it. In this vision, project managers need to take a broad view of their role and to evolve from 'doing things right' to 'doing the right things right'. This implies taking responsibility for the results of their projects, including the sustainability aspects of such results. Furthermore, the relationship between project management and sustainability is explored in academic studies (for example, by Gareis et al. 2009; Labuschagne and Brent 2006), as one of the necessary (future) developments in project management.

How does this attention for sustainability find its way to the shop floor? How is sustainability taken into account in project management processes, methodologies and competencies? Is it a point of concern there? If organizations 'put their money where their mouth is' on sustainability, it is inevitable that sustainability criteria and indicators will find their way into project management methodologies and practices in the very near future (Silvius et al. 2009).

This book explores the concept of sustainability and its application to projects and project management. It identifies the questions surrounding the integration of the concepts of sustainability within projects and project management and suggests answers to these questions. Or, when answers are not available, the book provides insights that may lead to more informed considerations. By doing this, the book aims to contribute to the further development of the project management profession.

For this reason, the book is, in the first place, aimed at project management professionals. It is this group that will be faced with an increasing demand for demonstrating sustainability in their work. But the responsibility for sustainability in projects also rests with general managers, project sponsors, project management office (PMO) leaders and other stakeholders in the context of projects. They represent the second group that this book is aimed at. The third and final group this book targets is educators and students in the field of project management and general management. They represent the future of projects and the project management profession, and should understand the developments that shape the project manager of the future: a project manager who takes responsibility for a sustainable future.

1.2 WHY SUSTAINABILITY IN PROJECTS AND PROJECT MANAGEMENT?

The rationale for studying the integration of sustainability is multiple. Firstly, the recognition that the current ways of exploiting the earth's natural resources are not sustainable, logically leads to the conclusion that we need to change the way we do things and/or the things we do. We need to change the way we use resources, produce products, make our preferences, share our wealth, and so on. And this change is inescapably related to innovation and projects. Sustainable development therefore needs projects to realize change. Secondly, this change is also related to the strategies of companies and other organizations. Organizations are moving towards sustainability and this influences the projects they undertake, but also the way they execute their projects. Sustainability therefore changes the profession of project management. This development provides a third reason for this book. Project management is developing into a 'true' profession and with this profession comes a professional responsibility. Perhaps even an ethical responsibility. Project managers should therefore take responsibility for integrating sustainability into their work.

However, ... if integrating considerations of sustainability in projects and project management is so logical and evident, would that not suggest that it is already being done? In other words, what's new? Isn't sustainability just another hype that suggests a radical innovation, but in reality is 'old hat'?

We aim to also address this question in this book. We hope to show the implications the concepts of sustainability may have on projects and project management and how these changes the profession of project management. Old hat? We'll reflect on that in Chapter 5.

1.3 OUR CORE MESSAGE

The core message of this book is a call to the project management profession. A call to understand that project managers can and should take responsibility for contributing to sustainable development by organizations and businesses. A call to stand up and be a specialist in change – sustainable change – which involves professional autonomy, professional responsibility and ethical behaviour.

In addition to being a call for action, the book is also a guide for this action. It provides practical guidance on how to integrate the concepts of sustainability in the way organisations execute, manage and govern projects.

This book is about project management as a profession. We support the opinion of Mary McKinlay that the further development of this profession requires project managers to take responsibility for sustainability. If project managers position themselves as subordinates of project sponsors and 'just do as they are told', without taking up a more autonomous responsibility, they cannot expect organizations and businesses to fully appreciate the significance of project management. Project managers should be the 'peers' of their project sponsors, specialized in organizational change and managing temporary organizations.

The book is also for individuals. Individuals that work in teams or with teams on projects. Individuals with knowledge, skills, attitudes, and personal values. Sustainability relates to personal values in terms of ethics, fairness, equality, responsibility and accountability. We hope to inspire these individuals to be aware of their personal values, and to apply these to the different roles they play in projects and change. We hope to inspire them to think about professionalism and professional responsibilities.

1.4   HOW TO USE THIS BOOK

This book 'translates' the concepts of sustainability into practically applicable insights on the integration of sustainability in projects and project management. Figure 1.1 shows this overall structure, from the exploration of the concepts of sustainability to practical steps to get started.

A brief introduction to the chapters:

Chapter 2, Sustainability in Business, provides an introduction into and an overview of the concepts of sustainability. It will be interesting to readers who want to understand these concepts and their relevance to doing business. The concepts of sustainability are summarized in six principles of sustainability that form the foundation for the analysis in the further chapters of the book.

Chapter 3, Sustainability and Projects, provides a brief introduction into the specific characteristics of projects and analyses the application of the principles of sustainability to projects. This chapter is directed at readers who want to understand the background and conceptual implications of the integration of the concepts for sustainability in projects and project management. The chapter concludes with a definition of sustainability in projects and project management and a checklist of sustainability aspects of projects.

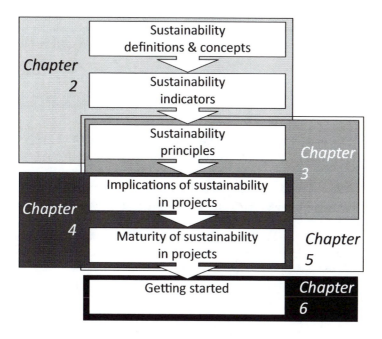

Figure 1.1   The overall structure of this book

Chapter 4, Incorporating Sustainability in Project Management, analyses the implications of the principles of sustainability, and the checklist of sustainability aspects. It discusses these implications initially at the level of the project (project management processes, maturity of sustainability integration); secondly at the level of the project manager (competences, ethical considerations); and thirdly at the level of the organization (project governance, organizational maturity). This is a practical chapter with specific analysis of the most commonly used project management standards, and suggestions and tools for the integration of sustainability in projects and project management. Therefore it is relevant to all readers.

Chapter 5, Reflection and Conclusion, is a more reflective and conclusive chapter. It is in this chapter that we will address the question posed in the previous section: Isn't sustainability just 'business as usual'? This chapter also addresses the relationship between sustainability, ethics and professionalism in project management. It is relevant for those who want to have a deeper understanding of what sustainability in projects and project management implies on a more philosophical level.

The final chapter, Chapter 6 Getting Started, provides practical guidelines to 'get started' with the integration of sustainability in projects and project management. This chapter is organized into the different roles surrounding projects. It is especially relevant for practitioners in these roles who wish to understand what they can do tomorrow to start taking responsibility for a more sustainable future.

REFERENCES

Association for Project Management (2006), *APM supports sustainability outlooks*, http://www.apm.org.uk/page.asp?categoryID=4, Accessed 14 June 2010.

Gareis, R., Huemann, M. and Martinuzzi, A. (2009), Relating sustainable development and project management, IRNOP IX, Berlin.

Gareis, R., Huemann, M. and Martinuzzi, A. (2011), What can project management learn from considering sustainability principles?, in *Project Perspectives*, Vol. XXXIII: 60–65.

Hedstrom, G., Poltorzycki, S. and Stroh, P. (1998), Sustainable development: the next generation, in *Sustainable Development: How Real, How Soon, and Who's Doing What?*, Prism Q4/98, 5–19, Arthur D. Little, Inc, Cambridge, MA.

Holliday, C. (2001), Sustainable growth, the DuPont way, in *Harvard Business Review*, 79(8), 129–134.

International Institute for Sustainable Development and Deloitte & Touche (1992), *Business Strategy for Sustainable Development: Leadership and Accountability for the 90s*, International Institute for Sustainable Development, Winnipeg.

Kennedy, A. (2000), *The End of Shareholder Value: Corporations at the Crossroads*, Basic Books, New York.

Labuschagne, C. and Brent, A.C. (2006), Social indicators for sustainable project and technology life cycle management in the process industry, in *International Journal of Life Cycle Assessment*, 11 (1): 3–15.

McKinlay, M. (2008), Where is Project Management running to...?, International Project Management Association, 22nd World Congress, Rome.

Silvius, A.J.G., Brink, J. van der and Köhler, A. (2009), Views on Sustainable Project Management, in Kähköhnen, K., Kazi, A. and Rekola, M. (Eds.), *Human Side of Projects in Modern Business*, 545- 556, IPMA Scientific Research Paper Series, Helsinki, Finland.

Silvius, A.J.G., Brink, J. van der and Köhler, A. (2010), The Impact of Sustainability on Project Management, Asia Pacific Research Conference on Project Management (APRPM), Melbourne.

Visser, W.T (2002), Sustainability reporting in South Africa, in *Corporate Environmental Strategy*, 9 (1): 79–85.

CHAPTER 2

SUSTAINABILITY IN BUSINESS
JULIA PLANKO AND GILBERT SILVIUS

This chapter introduces the concepts of sustainability and their application in the business world. It gives a brief overview of key developments of frameworks and organizations, which can be of help to companies who decide to operate more sustainably. Based on these concepts, six principles of sustainability will be defined. These principles play a leading role in the integration of sustainability in projects and project management that will be discussed in later chapters.

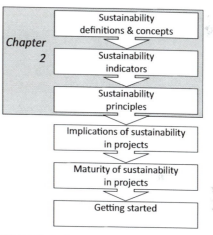

2.1   THE CONCEPTS OF SUSTAINABILITY

The term 'sustainability' has been interpreted and used in different ways. It basically means to operate in a way that may be sustained indefinitely, that is, to generate something without thereby depleting or destroying the necessary (re)sources. A prominent example is overfishing. If too many fish are caught in a certain period of time, fish stocks cannot reproduce themselves fast enough in order to fulfil the fishing needs of the consecutive year. If continued, it leads to extinction of the whole fish population, and the fishing industry cannot be sustained. Another example is the depletion of oil reserves. If we continue using petrol products at current level, oil reserves will be exhausted, and future generations will not be able to produce oil products at all, which will bring many industries to a halt, and reduce the availability of many oil based products that currently bring comfort to our lives.

The term can also be applied to people. If an employee works too hard over a long period of time, he or she may suffer from burn-out syndrome and may not be able to work again; he/she thus had an unsustainable working style.

Sustainable development has been defined as 'development that meets the needs of the present without compromising the ability of future generations to meet their own needs' (World Commission on Environment and Development 1987). This definition is widely used, and has often been interpreted as an appeal to only consume in such a way that we do not destroy too many resources, especially the environment, to allow future generations to have a good quality of life.

The core of many concepts of sustainability is that sustainable development implies a move towards economic prosperity, environmental protection and social equity. Developments in one dimension should not compromise the other dimensions. That is, economic growth should not be carried out to the detriment of the environment. Societal tensions often have negative impacts on the economy (for example, revolutions in Arab world in 2011) and on the environment (for example, garbage problems in Naples in 2010). Therefore, when a company, nongovernmental organization (NGO) or politician wants to implement a change, its impact on all three dimensions should be taken into account. A balance or harmony should be targetted between these three perspectives.

The three dimensions of sustainability, and their relations, are illustrated in Figure 2.1.

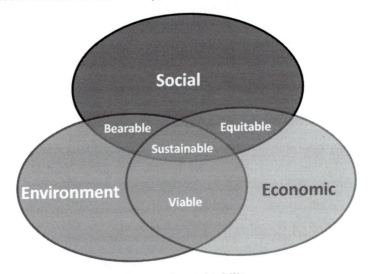

Figure 2.1   The three dimensions of sustainability

The term 'sustainable development' was first coined amongst activists and political leaders as a reaction to the exponential increase in resources used as a result of industrialization. The book *Limits to Growth* published by the Club of Rome in 1972 triggered this discussion on a global scale (Meadows et al. 1972). The authors simulated further exponential growth of world population, industrialization, pollution, food production and resource depletion and predicted its consequences on interactions between ecosystems and societies. Their conclusion was that if the world population and world economy continue to grow at their current speeds, our planet's natural resources will approach depletion. Future generations would need to live in an overpopulated world with a natural environment that would be destroyed to such a degree that it would unbearable.

The concerns about the sustainability of the earth and its resources, given the effects of economic development, led to the installation of the United Nation's (UN) 'World Commission on Development and Environment', also known as the Brundtlandt Commission. In 1987 the Brundtland Commission published the report 'Our Common Future' in which it sets out the principles of sustainable development. The acceptance of the Brundtland Report by the UN General Assembly placed sustainable development on the agenda of many governments.

However, not only governments worry about sustainable development. Many businesses and consumers also understand that we cannot continue to produce and consume the way we do now. Everyone who has ever tried to go on a diet or start a regular exercise regime knows that to understand something should be done and to act upon it are two different things. The warning signs are getting stronger and people are more informed. Tremendous oil spills, natural disasters, climate change and environment-related health problems are just a few of the more alarming signs that make the public aware that there is a problem. Whereas governmental regulations are essential for a change in status-quo, the driving force of this change can and perhaps should be businesses.

2.2   SUSTAINABILITY IN THE BUSINESS CONTEXT

The concept of sustainability is based on balance or harmony between the three dimensions: social equity, environmental protection and economic prosperity. In the business world these three dimensions are often called 'people, planet, profit' or PPP. The PPP concept implies that a company should take its decisions with consideration of people – its employees as well as other stakeholders and society – and the planet – that is, the environment – as well as profit. Since the primary goal of most companies is to generate shareholder value, the 'profit' dimension is well represented in business strategies and policies. Almost by definition, companies are organized with a strong orientation towards the economic perspective.

The environment and the social perspectives, however, are often less embedded in a company's strategies and practices. Hence, organizations who try to contribute to more sustainable development of our world should focus more on the reduction of the negative environmental and social impacts of its operations while maximizing positive environmental and social impacts.

Many of the larger companies of the world, like Unilever, General Electric and Walmart, have already accepted corporate sustainability as an integrated aspect of doing business. Strategies for integration involve appointing corporate sustainability officers, publishing sustainability reports and incorporating sustainability into their corporate communication strategies (Dyllick and Hockerts 2002). Clearly, like all business strategies, a sustainability strategy needs clear objectives and needs to be measured. In 1994, John Elkington coined the term 'triple bottom line' (TBL), in his book *Cannibals with Forks* (1997). TBL is based on the financial reporting term 'bottom line'. Elkington suggested that companies should not merely focus on their financial bottom line (net income or profit), but also on their achievements with regards to social and environmental aspects. A sustainable company should logically measure, monitor, document and report its impact on all three bottom lines. Financial or economic performance indicators may include: sales, return-on-investment, taxes paid, monetary flows and profit. Environmental or ecological indicators may include: air quality, water quality, energy usage and waste produced. Social performance indicators may include: jobs created, labour practices, community impacts, human rights and product responsibility (Savitz and Weber 2006).

Businesses can try to reduce their environmental impacts by changing their production processes. They can influence their suppliers by adapting procurement policies. They can have positive impacts on society by enabling their employees to improve their work–life balance. They can motivate staff to engage in projects which help society (for example, the company TNT sends out employees to help in the United Nation's World Food Program). They can compensate with projects for environmental damage they have done. They can help customers behave more sustainably by designing sustainable products. Depending on the sector they are in, they can lease out instead of sell their products and develop products that are durable for a longer term. They can train their employees, make them more aware on the topic and encourage them to always consider all three dimensions (PPP) when taking decisions.

A number of organizations have developed frameworks that help companies in this transformation and many books have been written on this topic. The following section gives a brief overview of important frameworks which companies can use when embarking on their journey.

- As a response to businesses' growing interest and the increasing number of sustainability-related institutions and frameworks, the International Organization for Standardization (ISO) launched ISO 26000, a comprehensive guideline on social responsibility, to help companies introduce more sustainable practices. ISO 26000 is a guideline on social responsibility. It is designed for all types of organizations, that is, also for NGOs and governments.

ISO 26000 defines social responsibility as:

Social responsibility is responsibility of an organization for the impacts of its decisions and activities on society and the environment, through transparent and ethical behaviour that
- contributes to sustainable development, including health and the welfare of society;
- takes into account the expectations of stakeholders;
- is in compliance with applicable law and consistent with international norms of behaviour;
and
- is integrated throughout the organization and practiced in its relationships.
Note 1: Activities include products, services and processes.
Note 2: Relationships refer to an organization's activities within its sphere of influence.

A company can consult or adopt ISO 26000 to learn more about social responsibility and as guidance for implementing sustainability principles and practices into their business processes, strategies, procedures, systems and organizational structures. Annex A gives an overview of what ISO 26000 considers to be the main areas of interest for companies who aspire to be more sustainable. It summarizes seven social responsibility 'core subjects'. These core subjects are further broken down into 'issues', specific themes or activities a company should work on in order to contribute to sustainable development.

- Another frequently used ISO standard is ISO 14001. This standard describes the core set of processes and practices for designing and implementing an effective environmental management system in an organization, similar to the ISO 9000 series of standards on quality management. ISO 14001 is a normative standard, unlike ISO 26000, which is a guideline. This means that organizations can certify themselves as being compliant with ISO 14001. ISO 14001 and ISO 26000 do not include or imply each other. They can be used independently of, but also complimentary to each other.

- The Global Reporting Initiative (GRI) is a non-profit organization that pioneered the world's most widely used sustainability reporting framework. Companies can use the framework to indicate to shareholders and consumers their economic, social and environmental performance. The most current version is called the G3 Guidelines and is a available at no charge. GRI's objective is to facilitate sustainability reporting for companies and thereby stimulate them to operate more sustainably. The framework is made up of various indicators, from which companies can select the ones they find most suitable for their own operations (Global Reporting Initiative 2010). Table 2.1 shows an overview of the GRI criteria.

Table 2.1   Overview of indicators in the Sustainability Reporting Guidelines

Economic Sustainability	Direct Economic Performance
	Market Presence
	Indirect Economic Impact
Environmental Sustainability	Materials
	Energy
	Water
	Biodiversity
	Emissions, effluents and waste
	Products and services
	Compliance
	Transport
	Overall
Social Sustainability	Labour Practices and Decent Work
	Human Rights
	Society
	Product Responsibility
	Ethics

- The UN Global Compact is a framework of ten universally accepted principles, developed by the UN and a number of large corporations. It covers the areas of human rights, labour, environment and anti-corruption. Participating companies agree to comply with these principles. They can use the provided framework as a platform for disclosure. This initiative has been created because the UN realized that businesses are primary drivers for globalization and can help ensure long-term value creation that can bring benefit to economies and societies all over the globe. In the absence of global regulations, this voluntary code of conduct has been developed, hoping to stimulate companies to more sustainable business practices. (United Nations Global Compact 2010).

- The Natural Step Framework is a holistic framework which helps organizations to integrate sustainability principles into their business strategies. It provides a tool for them to develop a shared vision, shared identity and shared goals among departments and along supply chains.

Foundation of the Natural Step Framework is the principle that a company should try to reduce its negative impacts on the biosphere while enabling humans to fulfil their needs. It stimulates companies to re-think production processes and product design and to find innovative alternatives for achieving their business goals. The participatory framework provides a good basis for both awareness raising as well as strategy development.

2.3   DEVELOPMENT OF SUSTAINABILITY IN BUSINESS: FROM LESS BAD TO GOOD

As mentioned earlier, many companies have already accepted some level of responsibility for sustainable development as part of being in business. Stimulated by government regulations, the voice of pressure groups or perhaps by their own beliefs and values, they are implementing more sustainable business practices in their organizations. The examples are numerous, from more 'fair' production of raw materials, to more efficient use of energy, to tighter travel policies, to recycling of end-of-life products, and so on. However, the level to which organizations integrate sustainable practices into their operations varies widely.

In his book *The Next Sustainability Wave*, Bob Willard (2005) describes five 'sustainability stages' for a company (see Figure 2.2). The stages move from reactive to proactive and describe to what extent a company is committed to sustainability principles. The first stage is when companies fail even to comply with prevailing regulations. They are opportunistic and not engaged with the concept of sustainability. When a company complies with all environmental and social regulations it moves up to stage 2, 'compliance'. In stage 3, 'beyond compliance', a company starts to not only react on regulations, but it starts introducing sustainability activities. Yet, these activities are not concerted but are

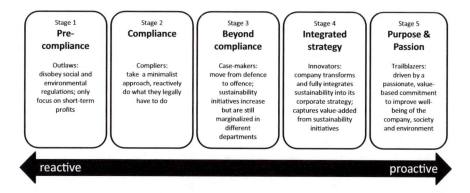

Figure 2.2   The stages of sustainability
Source: Adapted from Willard, 'The Next Sustainability Wave', 2005

carried out in different departments. Companies who understand the importance of sustainability and the value-added they can gain from sustainable activities, for example, energy-efficient production or eco-friendly products, and integrate sustainability into their corporate strategy are in stage 4, 'integrated strategy'. The highest level, stage 5, 'purpose and passion', is attained when companies are not just driven by profits but also by a sense of responsibility to improve society and environment and to contribute to a better world.

The stages of sustainability model illustrates how companies can move from a reactive approach to a proactive one. The difference between the stages 3 and 4 is especially important here. This step distinguishes companies that merely try to be 'less bad', by trying to minimize the negative effects of their business, from companies that try to 'do good', by integrating sustainability considerations in their core business itself. This final stage, seeing sustainability as an integrated part of the business, is rapidly gaining momentum. It is the mind-shift from considering sustainability as a 'threat' and a 'cost' to considering it as a (business) opportunity.

2.4   THE BUSINESS CASE OF SUSTAINABILITY

Companies can play a key role in sustainable development. But why should they? What are the commercial reasons why managers should work on sustainability-transformations? What about the costs involved? Can sustainable companies still be profitable? Do the benefits for companies outweigh the costs?

The introduction of more sustainable business practices may require high initial investments in improved resources, revised business processes, new equipment and machines. Projects may also need to be undertaken to guide employees towards more sustainable work practices, or to develop more sustainable products and services. These investments usually take time to generate revenues. So, in the short term, the investments will probably outweigh the immediate benefits. In the longer run, the benefits may be more supportive to the business case. But what are the benefits? Financial or non-financial?

How does sustainability add to business goals or performance?

First of all, there are financial considerations. Sustainable practices can *decrease costs*. Eco-efficiency leads to costs savings. Processes that use less materials, energy and water and generate less waste are less costly for companies. Using more durable products and processes also increases the lifespan of products and thereby increases shareholder value (Figge 2001). Yet, sustainability is not just about decreasing costs. It can also be about *increased revenues*. General Electric's (GE) 'Ecomagination' initiative, which aims at developing and providing products that generate significant environmental and economic performance, provides proof for this. In 2010, GE decided to increase its commitment to the programme by investing US$10 billion in research and development over the next five years, effectively doubling the investment it had made in the previous five years. The reasoning for GE was simple: ecomagination is a cash cow, generating US$70 billion in revenue since its inception in 2005. GE's Ecomagination programme shows that there is a growing demand for sustainable products. It also shows that investing in research and development for more sustainable products can mean *staying ahead of competition*. And although any 'competitive advantage' effects may not hold in the longer term, for example Toyota's hybrid car concept that was copied by Honda and other manufacturers, developing sophisticated sustainable products may still lead to a temporary increase in revenues.

Next to foreseeable costs and revenues, *risk* may also be a consideration in adapting more sustainable business practices. For example, the 2010 British Petroleum (BP) oil spill disaster in the Gulf of Mexico resulted in clean-up costs, legal fees and government fines of over US$40 billion, not to mention the damage caused to BP's image and brand. And although this may seem like a cold-hearted calculation on one of the largest environmental disasters of the past few decades, the example shows the financial risk of not considering the business aspects of sustainability seriously enough.

However, financial considerations are only part of the business case. Quite likely, there will also be non-financial benefits to sustainability. For example, *shareholder's expectations* can have an impact. The topic of sustainability is becoming an increasingly important topic in boardrooms, and shareholders are demanding more public accountability of companies (Figge 2001). Banks and other investors are showing an increased interest in the sustainability policies of the organizations they are investing in. They increasingly see environmental and social performance as key to a company's future business success (Bergius 2008). For example, at VanderLande Industries, a Dutch producer of industrial conveyer belts, the suggestions of their main shareholder led to a unique commitment to sustainability as an integrated element of their products and their business strategy.

Some banks specifically mention sustainability as one of their main criteria for investments and, for example, pension funds feel an increasing pressure from their members to be transparent about the kind of organizations in which they invest their money. Therefore, adopting more sustainable business practices may provide easier *access to capital*.

Last but not least, with a new generation of potential employees who have learned about environmental issues in school and who are increasingly sensitive to the topic, sustainability can become a crucial issue in *attracting* the most *qualified personnel*. Sustainable practices can also lead to higher work motivation. For example, BMW's introduction of equal wages for white and black workers in South Africa, and the improvement in their living conditions and medical care, has drastically improved employee motivation and competence. These factories are now amongst the most productive automotive plants in the world.

Although the potential benefits of sustainability may be plentiful, the extent to which they compensate for the investment involved, needs to be considered case-by-case. Heel et al. (2001) argue that sustainability can also be seen as a driver of financial results, through corporate reputation and brand value. The main obstacle, however, is the time gap between the investment and levered benefits.

An interesting study by the Aberdeen Group showed that companies that take on a more proactive approach to sustainability – for example, as illustrated by Willard's stages of sustainability model discussed in section 2.3 – and integrate sustainability in their business strategy, actually gain from this in terms of a reduced carbon footprint, lower costs for energy, facilities, paper and transportation, and also have a mean of 16 per cent customer retention as compared to 4–5 per cent of the rest of the industry (Senxian and Jutras 2009). Futhermore, they found that the 30 per cent 'laggards' in adopting more sustainable business practices are actually confronted with higher costs (Table 2.2).

Table 2.2   Performance of different maturity classes on sustainability

Sustainability Performance Indicators	*Mean Class Performance*		
	Best-in-Class	***Industry Average***	***Laggards***
Carbon footprint	-/- 9%	-/- 6%	+5%
Energy costs	-/- 6%	+ 4%	+ 18%
Facilities costs	-/- 7%	-/- 1%	+ 18%
Paper costs	-/- 10%	-/- 5%	+ 12%
Transportation/ Logistics costs	-/- 7%	-/- 1%	+ 19%
Customer retention	+ 16%	+ 5%	+ 4%
	Top 20% of aggregate performance scorers	Middle 50% of aggregate performance scorers	Bottom 30% of aggregate performance scorers

Source: Adapted from Senxian and Jutras 2009

Although these results do not prove the business case for any individual investment in sustainability, they do provide indications that a more pro-active approach to sustainability pays off.

2.5   PRINCIPLES OF SUSTAINABILITY

Based on the concepts, standards and literature on sustainability, a number of key elements, or principles of sustainability, can be derived. For example, Dyllick and Hockerts (2002) identify three 'key elements of corporate sustainability': integrating the economic, ecological and social aspects into the firm's strategy; integrating short-term and long-term aspects; and consuming the income and not the capital. Gareis et al. define sustainability with the following principles (Gareis et al. 2011): economic, social and ecologic orientation; short-, mid- and long-term orientation; local, regional and global orientation; value orientation. The ISO 26000 mentions accountability, transparency, ethical behaviour, respect for stakeholders' interests, respect for rule of law, respect for international norms of behaviour and respect for human rights as 'principles' of sustainability. After considering these sets of elements or principles we concluded six 'principles of sustainability' that will act as guiding principles in the integration of the concepts of sustainability in projects and project management.

The six principles of sustainability are:

1. *Sustainability is about balancing or harmonizing social, environmental and economic interests*
   In order to contribute to sustainable development, a company should satisfy all 'three pillars' of sustainability as illustrated in Figure 2.1: social, environment and economic. The dimensions are interrelated, that is, they influence each other in various ways. Based on the thinking in Willard's model of stages of sustainability, organizations, organizations can have a reactive approach to this principle and try to 'balance' social, economic and environmental aspects by trading off the negative effects of doing business for a somewhat lower profit. For example, compensating CO_2 emissions by planting new trees or compensating unhealthy work pressure with higher salaries. A more proactive approach to sustainability looks at how organizations create a 'harmony' of social, environmental and economic aspects in their activities. This approach is not about compensating bad effects, but about creating good effects.

2. *Sustainability is about both short-term and long-term orientation*
   A sustainable company should consider both short-term and long-term consequences of their actions, and not only focus on short-term gains. Firms listed on the stock market may especially be tempted to focus on short-term gains, trying to increase performance from quarterly report to quarterly report, and thereby loosing long-term vision.
   This principle focuses the attention to the full lifespan of the matter at hand. An important notion with regards to this principle is that the economical perspective, because of discounting of future cash flows, values short-term effects more than long-term effects. In economic theory, an immediate cash flow holds more value than a future cash flow, thereby emphasizing the value

of short-term benefits. However, social impacts or environmental degradation because of business decisions, may not occur before the long-term.

3. *Sustainability is about local and global orientation*

   The increasing globalization of economies effects the geographical area that organizations influence. Intentionally or not, many organizations are influenced by international stakeholders whether these are competitors, suppliers or (potential) customers. The behaviour and actions of organizations therefore have an effect on economical, social and environmental aspects, both locally and globally. 'In order to efficiently address these nested and interlinked processes sustainable development has to be a coordinated effort playing out across several levels, ranging from the global to the regional and the local' (Gareis et al. 2011).

4. *Sustainability is about consuming income, not capital*

   Sustainability implies that nature's ability to produce or generate resources or energy remains intact. This means that the source and sink functions of the environment should not be degraded. Therefore, the extraction of renewable resources should not exceed the rate at which they are renewed, and the absorptive capacity of the environment to assimilate waste should not be exceeded.' (Gilbert et al. 1996). The economic equivalent of this principle is common knowledge in finance and business. Financial managers know that a company which does not use its income to pay for its costs, but instead uses its capital, will soon be insolvent. The principle may also be applied to the social perspectives. Organizations should also not 'deplete' people's ability to produce or generate labour or knowledge by physical or mental exhaustion. In order to be sustainable, companies have to manage not only their economic capital, but also their social and environmental capital.

5. *Sustainability is about transparency and accountability*

   The principle of transparency implies that an organization is open about its policies, decisions and actions, including the environmental and social effects of those actions and policies. This implies that organizations provide timely, clear and relevant information to their stakeholders so that the stakeholders can evaluate the organization's actions and can address potential issues with these actions. The principle of accountability is logically connected to this proactive stakeholder engagement. This principle implies that an organization is responsible for its policies, decisions and actions and the effect of them on environment and society. The principle also implies that an organization accepts this responsibility and is willing to be held accountable for these policies, decisions and actions.

6. *Sustainability is about personal values and ethics*

   As discussed earlier, a key element of sustainability is change: change towards more sustainable (business) practices. As argued by Robinson (2004) and Martens (2006), sustainable development is inevitably a normative concept, reflecting values and ethical considerations of society.

Part of the change needed for more a sustainable development will therefore also be the implicit or explicit set of values that individuals, business professionals or consumers have and that influence or lead behaviour. GRI Deputy Director, Nelmara Arbex, puts it simple and clear: 'In order to change the way we DO things, we need to change the way we VIEW things.'

These six sustainability principles provide guidance for the analysis of the impact of the concepts of sustainability in projects and project management in the following chapters.

REFERENCES

Dyllick, T. and Hockerts, K. (2002), Beyond the business case for corporate sustainability, in *Business Strategy and the Environment*, 11 (2): 130–141.

Elkington, J. (1997), *Cannibals with Forks: The Triple Bottom Line of 21st Century Business*, Capstone Publishing Ltc. Oxford.

Figge, F. (2001), *Keine Nachhaltigkeit ohne Oekonomischen Erfolg. Kein Oekonomischer Erfolg ohne Nachhaltigkeit [No Sustainability without Economic Success. No Economic Success without Sustainability]*, PriceWaterhouseCoopers, Frankfurt.

Gareis, R., Huemann, M. and Martinuzzi, A. (2011), What can project management learn from considering sustainability principles?, in *Project Perspectives*, Vol. XXXIII: 60–65.

Gilbert, R., Stevenson, D., Girardet, H. and Stern, R. (Eds), (1996), *Making Cities Work: The Role of Local Authorities in the Urban Environment*, Earthscan Publications Ltd, Oxford.

Global Reporting Initiative (2010), *Global Reporting Initiative*, Retrieved 22 October, 2010. Available at: http://www.globalreporting.org.

Heel, O.D., Elkington, J., Fennell, S. and Dijk, F. van (2001), *Buried Treasure*, SustainAbility, London.

Martens, P. (2006), Sustainability: science or fiction? in *Sustainability: Science, Practice, and Policy*, 2 (1): 36–41.

Meadows, D.H., Meadows, D.L., Randers, J. and Behrens, W.W. (1972), *The Limits to Growth*, Universe Books, New York.

Robinson, J. (2004), Squaring the circle? Some thoughts on the idea of sustainable development, in *Ecological Economics*, 48, 369–384.

Savitz, A.W. and Weber, K. (2006), *The Triple Bottom Line*, Jossey-Bass, San Francisco.

Senxian, J. and Jutras, C. (2009), *The ROI of Sustainability: Making the Business Case*, Aberdeen Group, Theale.

United Nations Global Compact (2010), *UN Global Compact*, Retrieved 22 October, 2010. Available at: http://www.unglobalcompact.org/AboutTheGC/index.html.

Willard, B. (2005), *The Next Sustainability Wave: Building Boardroom Buy-in*, New Society Publishers, Gabriola Island.

World Commission on Environment and Development (1987), *Our Common Future*, Oxford University Press, Great Britain.

CHAPTER 3

SUSTAINABILITY AND PROJECTS
GILBERT SILVIUS AND RON SCHIPPER

This chapter reviews the nature of projects and project management and its relationship with sustainability. After a brief overview of the history and expected future developments of project management, it will become clear that project management of the future will need to be more contextually oriented than it is today. Since the preceding chapter showed that companies are starting to integrate sustainability in the way they do business and manage their organizations, sustainability is getting more and more relevant to the resulting projects and necessary project management.

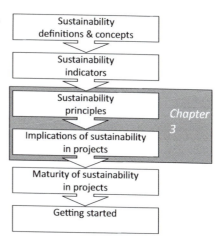

In this chapter we will also present an overview of the existing views on the integration of sustainability in projects and project management and explore what the implications of the principles of sustainability may entail for the way projects are undertaken and managed.

3.1   PROJECTS AND PROJECT MANAGEMENT

Although it should be expected that in ancient Egypt some form of project management was practised in the construction of unique artefacts like the pyramids, it took until the 1950s to see the first developments of project management as a discipline (Turner et al. 2010). The most important forerunner of this development in the United States was Henry Gantt. From his study of the management of Navy ship building, he developed, early last century, a scheduling technique for projects that today is still considered an essential part of project management theory and practice: the Gantt chart or bar chart. The 1950s, however, brought the development of mathematical project scheduling models like the 'Program

Evaluation and Review Technique' (PERT) and the 'Critical Path Method' (CPM). These techniques marked the take-off of project management as a discipline.

Until the 1970s, the traditional fields of application of project management were construction, engineering and defence. Project management methods and theory were dominated by a 'hard' objective and scientific paradigm. In this 'hard' paradigm, the world is seen as an objective reality and 'systems are mechanistic processes, with stable, or predictably' outcomes (Crawford and Pollack 2004). This background may still be reflected in the definition of a project as 'a temporary endeavour undertaken to create a unique product, service or result' (Project Management Institute 2008).

However, with the emergence of information technology in later decennia, the applications of project management spread into all industries and contexts. And with this spread, the nature of project management changed. From a 'hard' mathematical technique to optimize planning, construction and production processes it developed into an instrument to manage organizational change (Gareis 2010). Therefore, a different, 'soft' paradigm to project management emerged in which projects evolve around social interaction rather than around mathematical optimization, and in which goals, methods, expectations, solutions, outcomes and success of projects are less manageable or predictable (Crawford and Pollack 2004).

This change in paradigm is illustrated by a study of Kloppenborg and Opfer (2002) into the project management research published in English since 1960. An annotated bibliography was created of 3,554 articles, papers, dissertations and government research reports. The study identified a distinct shift in topics of interest over the decades. In the 1960s, most research focused on large, defence-related projects. In the 1970s, the research focused on cost and schedule control, performance measurement, work breakdown structures and life cycle management. While cost/ schedule control remained a topic of major research interest during the 1980s, research started to include team building, quality and knowledge management-related topics. The 1990s saw a further increase in Human Resource (HR) and organization-related topics such as organizational change, team development and leadership, as well as a focus on risk management. So the nature of project management evolved from the application of mathematical techniques, to the organization of structured processes and the management of organizational change.

The 'organizational change' view on projects caused a need for further development of the project management profession. Traditional project management techniques and theory originated from systems management or cybernetics theory. With the scope of projects however, shifting from constructing or creating 'things' to organizational change, the complexity of the 'system' at hand increased. Based on Boulding's Classification of Systems (Annex B), it could be argued that project management as an instrument of change needed to develop from the management of a level 3 'thermostat' system, with predictable input–action–output relationships, to a level 8 'social organizations' system, with far less predictable outcomes.

In line with the development of projects and project management described above, this book views project management as the management of project-organized change in organizations, policies, products, services, processes, assets and/ or resources, in or between organizations. These project-organized changes, or simply projects, are characterized by:

- a temporary nature or temporary organization;
- most often across organizational structures and boundaries;
- a defined deliverable or result, logically or preferably linked to the organization's strategy or goals;
- specified resources and budget.

In this definition, projects are, as temporary organisations, related to non-temporary 'permanent' organisations, and realize changes that benefit the strategy or goals of such organisations.

The permanent organisations utilise resources and assets in their operational business processes to deliver products and/or servicesto their customers, creating benefits, and ultimately performance (for example, profit, market share, employment, return on capital and so on), to the organizations and their stakeholders. Their activities are based on goals or ambitions that are developed or set in a strategic management process. Figure 3.1 provides a schematic overview of this relationship between goal setting, the utilization of assets and resources, operations, benefits and performance.

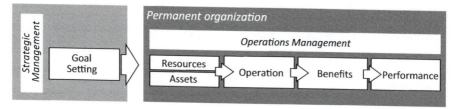

Figure 3.1   Schematic overview of relationships within the permanent organization

However, the strategic management of the organization doesn't just include the setting of goals. It also includes evaluating the business performance of the organization against these goals. If the performance is satisfactory, the operations may continue. But if the performance is unsatisfactory, because of lack of performance or because of changing goals, there may be reason to change something in the organization. In such case, a temporary arrangement or organization, in the form of a project, is commonly used to create this change. The change may concern the resources, assets or business processes of the permanent organization, but also the products/services rendered or the internal policies and procedures. The selection of the 'right' changes for the organization is usually part of a process called 'portfolio management'. Figure 3.2 illustrates this relationship between projects as temporary organizations and the permanent organization.

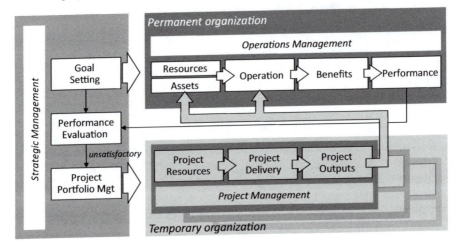

Figure 3.2   Projects as temporary organizations that deliver changes to the permanent organization

The evolution of the nature of project management caused the traditional 'iron triangle' of project management (scope/quality, time and cost/budget) to be increasingly inadequate. Managing social organization system-level change needs more holistic concepts of project management. One of these concepts is 'complex project management'. Complex project management arose from the application of complexity theory to project management. A complex system is a system formed out of many interacting components whose behaviour may be partially dependent and partially independent. Projects may be complex systems (Dombkins 2006), not only because they deal with complex technological issues but also because they deal with the wider organizational factors largely beyond the project manager's control (Xia and Lee 2004).

Complex project management does not abolish or discard the project management techniques such as CPM, PERT and the traditional 'iron triangle' of scope/quality, time and cost/budget. It does, however, acknowledge that these techniques do not model the reality of the uncertainty of the project environment well (Rand 2000). Its solution for the management of complex projects is therefore not a set of techniques, but a more holistic approach in which human relations play a central role.

Whether this development of project management can be considered complete is open for discussion, but the limitations of the 'mechanical' approach to project management are widely recognized. However, more holistic concepts, like complex project management, do not seem to have reached maturity or general acceptance yet.

Australian professor, and Director of the Complex Program Group, David Dombkins, pictures the development of the project management profession as presented in Figure 3.3. His representation positions a number of important highlights in the development of project management on a development curve that depicts the rate of change of the profession.

Dompkins envisions a new relatively stable period to arise from the rollercoaster of developments the profession has experienced in the five last decades. However, given the economic turbulence of the past years, the political changes that may result from that, the ongoing technological progress and the need for a more balanced approach to economic, environmental and social development, a relatively stable future seems unlikely. The changes that organizations face in the next 20 years may be just as, or even more, challenging, than the changes in the last 20 years. It is the responsibility of project management as a profession to contribute to our ability to handle these challenges. We should therefore expect an ongoing, and perhaps even accelerated, development of the project management profession in the near future. But how may the profession develop?

3.2   THE FUTURE OF PROJECT MANAGEMENT

In the introduction, we mentioned Mary McKinlay's keynote speech at the 2008 International Project Management Association (IPMA) World Congress as one of the motivations for this book. Her vision that 'the further development of the project management profession requires project managers to take responsibility for sustainability' (McKinlay 2008) connects sustainability to the future of project management. But why does this make sense? What is the logic of this connection?

The future of project management is frequently addressed in visionary papers and presentations. At the 15th IPMA World Congress on Project Management in 2000, Gorrino-Arriaga and Eraso reflected on the future on project management. Based on a substantial literature review, they identified a growing need for project managers to understand the business context of their projects, and discussed new processes and competencies that will be needed to cope with this (Gorrino-Arriaga and Eraso 2000). In line with this development, Hartman (2001) states that business competitiveness today requires constant change. The ability of organizations to realize change is severely limited by the static nature of the organizational model itself. Project management is therefore being used as the preferred way

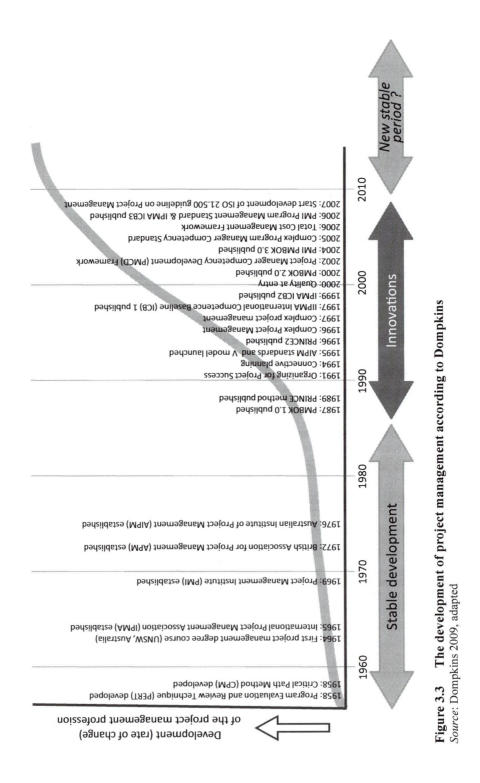

Figure 3.3   The development of project management according to Dompkins
Source: Dompkins 2009, adapted

The following content appears within the figure:

New stable period ?

Innovations

Stable development

Development (rate of change) of the project management profession

1960 1970 1980 1990 2000 2010

1958: Program Evaluation and Review Technique (PERT) developed
1958: Critical Path Method (CPM) developed
1964: First project management degree course (UNSW, Australia)
1965: International Project Management Association (IPMA) established
1969: Project Management Institute (PMI) established
1972: British Association for Project Management (APM) established
1976: Australian Institute of Project Management (AIPM) established
1987: PMBOK 1.0 published
1989: PRINCE method published
1991: Organizing for Project Success
1994: Connective planning
1995: AIPM standards and V model launched
1996: PRINCE2 published
1996: Complex Project Management
1997: Complex project management
1997: IPMA International Competence Baseline (ICB) 1 published
1999: IPMA ICB2 published
2000: Quality at entry
2000: PMBOK 2.0 published
2002: Project Manager Competency Development (PMCD) Framework
2004: PMI PMBOK 3.0 published
2005: Complex Program Manager Competency Standard
2006: Total Cost Management Framework
2006: PMI Program Management Standard & IPMA ICB3 published
2007: Start development of ISO 21.500 guideline on Project Management

of implementing technological and other change. Hartman argues that the next step for project management lies in the needs of organizations tomorrow. Since Chapter 2 showed that the importance of integrating sustainability in business practice is growing, and many organizations have already embraced sustainability as a principle of doing business, this increased orientation on the business context of projects lead projects towards also adressing sustainability in the way projects are executed and managed.

An increased business orientation can also be found with Heerkens (2001) when he speaks about the 'business savvy' project manager. He argues that the project manager's role is viewed as largely that of a producer of technical solutions, bound by the triple constraints of the iron triangle, but that this view will need to change in the future. Organizations wishing to thrive, or perhaps just survive, in the turbulent, competitive, and rapidly changing environment in the future will need to give their project managers the opportunity and the latitude to act in a more entrepreneurial fashion. And for the most part, the measure of the project manager's entrepreneurial expertise will be evaluated by their ability to demonstrate and apply sound business judgement. Of course, the project managers of the future will still have to produce deliverables that function as expected and satisfy requirements. But the project managers of the future will also be expected to do this in a way that ensures a positive contribution to the 'bottom line' performance of the organization. Metrics such as profitability, cash flow, life cycle cost, and strategic alignment will gradually begin to eclipse scope, cost and schedule as the essential measures of project success. And although more than 10 years old, Heerkens' views may still be somewhat foreign to many of today's project managers, because the business success of a project is ordinarily not defined as one of their key areas of responsibility. Foti (2001) takes the relationship between project management and business one step further and argues that the two may become synonymous with each other. He asks himself whether project management will become the standard operating procedure for smart and dynamic companies.

The growing importance of the contextual orientation of projects also changes the way projects are performed and managed. For example, Woollett (2000) argues that the project management practised is still stuck in the 1980s and out of date. Projects today exist in an environment of globalization, virtual teams, mobility and momentous advances in technology and communication. In line with this increased dynamic, Barnes (2002) foresees the change from the predictable model of project management to the unpredictable. Many projects are still based on the assumption that a project can be carried out in accordance with the original plan if the planning is good enough. This 'stable' model of project management, however, is far from reality. The rejection of the stable model has begun, but still has a way to go. The unstable view on project management says that the project sponsor, but also a real project manager, will change the plan whenever a better one can be established. Jaafari (1998) identified an increasing trend to require project managers to accept at least partial responsibility for the results their project realizes and the commercial success of these results. Also he believes that the traditional models of project management are increasingly inadequate in the highly turbulent and technology dependent world and makes a case for a fundamental shift in the preparation of the next generation of project managers. This 'next generation' should be proficient in the core processes and technology of the organization's operation. Next to this business orientation, these project managers should also be capable of integrating technology, be IT literate and be capable of operating within a concurrent engineering/construction environment. Preparation of such professionals will require skills in systems engineering and knowledge management.

Overlooking these studies, the expected (future) development of project management can be summarized as follows.

- First of all, most authors seem to share the understanding mentioned in section 1.1 that competitiveness of organizations, now and in the future, requires constant change of these companies. This change will mostly be organized in projects and therefore project management becomes a critical competence of any organization.
- This development requires project management in the future to be more oriented on the business context of a project as well as on the (triple) constraints of a project itself. The necessary shift in orientation results from an increasingly dynamic and turbulent environment which does not allow for fixed goals over any realistic period of time. Aligned with the development described above, 'soft values' and leadership tend to become more important compared to technical project management skills.
- Thirdly and finally the authors seem to agree on the observation that 'the sometimes accidental job of project manager is developing into a profession'. At this moment project management may still be an emerging profession, but the increasing profile, proliferation of courses and trainings, sophistication of software applications, maturity of professional associations and rising academic interest in project management adds to this professionalization.

An interesting study by Ingason and Jonasson (2006) on the development of project management from 2006 to 2020 also provides a comprehensive overview of these developments. Figure 3.4 summarizes this view of their Icelandic think-tank.

Although none of the studies discussed in this section actually mentions sustainability as one of the future developments in project management, the increased contextual orientation identified by several authors actually requires the project management profession to consider trends and developments in business and society in general. And since organizations show a growing concern for the sustainability issues connected to their businesses, the developments summarizing the future of project

2006	2020
Mechanical project management	Human project management
Single project management	Portfolio project management
Less project management mindset	More project management mindset
Hard value measurement	Soft value measurement
Short term thinking	Long term thinking
Narrow economy	Wide economy
Less contextual	Contextual
Focus on results	Focus on path
Emphasis on safety	Emphasis on strategy
Corporate growth	Personal growth
Specialization	Renaissance
Measuring what is measurable	Measuring what is important to measurable
Non-transparent	Transparent
Self-preservation	Self-transformation
Less-abundant living	Abundant living
Do things right	Do the right things
Defensive thinking and practices	Strategic thinking and practices
Process thinking	Transformative thinking
Focus on objectives	Focus on freedom
Linear thinking	Non-linear thinking
Micro management	Macro management

Figure 3.4   The future development of project management: views of an Icelandic think-tank

Source: Ingason and Jonasson, 2006

management support the argument for considering sustainability in projects and project management.

3.3   VIEWS ON SUSTAINABILITY IN PROJECTS AND PROJECT MANAGEMENT

The concerns about sustainability illustrated in Chapter 2 indicate that the current way of producing, organizing, consuming and living may, or will, have negative effects on the future. In short, our current way of doing 'things' is not sustainable. Therefore, 'things' (products, processes, materials, resources, our behaviour), but also 'thoughts' (beliefs, values) need to change. Elaborating on the view of projects as instruments of change, it is evident that a (more) sustainable society requires projects to realize sustainable change. In fact, this connection between sustainability and projects was already established by the World Commission on Environment and Development (1987). However, Eid, in his study on the integration of sustainable development into project management processes, concludes two decades later that the standards for project management 'fail to seriously address the sustainability agenda' (Eid 2009). Given the nature of projects as temporary organizations, this conclusion may not be surprising. Projects and sustainable development are probably not 'natural friends'. Table 3.1 illustrates some of the 'natural' differences in the characteristics of the two concepts.

Table 3.1   The contrast between the concepts of sustainable development and projects

Sustainable Development	Project Management
Long-term + short-term oriented ⟵——————⟶	Short-term oriented
In the interest of this generation ⟵——————⟶ and future generations	In the interest of Sponsor/Stakeholders
Life cycle oriented ⟵——————⟶	Deliverable/result oriented
People, Planet, Profit ⟵——————⟶	Scope, Time, Budget
Increasing complexity ⟵——————⟶	Reduced complexity

The relationship between sustainability and project management is still an emerging field of study (Gareis et al. 2009). Some first studies and ideas were published in recent years. And although the studies differ in approach and depth, a few conclusions can be drawn.

Conclusion 1: Sustainability is relevant to project managers

As stated in the introduction to this book, Association for Project Management's (APM) (past-)chairman, Tom Taylor, was one of the first to suggest that the project management community familiarize themselves with the issues of sustainability, recognizing that more should be done to contribute to a more sustainable society. This appeal was the output of a small working party in APM that recognized that project managers were not well equipped to make a contribution to sustainable development and decided to investigate this issue.

At the 2008 European conference of the Project Management Institute (PMI), Jennifer Russell elaborated on what corporate social responsibility means for project managers. She pointed out that a project manager, being in the frontline of new or changed activities within an organization, is perfectly positioned to influence the organization's operations towards greater sustainability. Russell also argued that this position is not without responsibility, both for the organization and the project manager. She concluded that 'Corporate social responsibility is too big an issue to leave to someone else to address' (Russell 2008).

Conclusion 2: Integrating sustainability stretches the system boundaries of project management

In some of the first publications on sustainability and project management, Carin Labuschagne and Alan Brent of the University of Pretoria link the principles of sustainable development to project life cycle management in the manufacturing industry (Labuschagne and Brent 2006). They suggest that the future-orientation of sustainability implies that the full life cycle of a project, from its conception to its disposal, should be considered. Elaborating on this life cycle view, Labuschagne and Brent argue that when considering sustainability in project management, not just the total life cycle of the project (for example, initiation–development–execution–testing–launch) should be taken into account, but also of the 'result' the project produces, being a change in products, assets, systems, processes or behaviour. This result, in their words: the 'asset', should also be considered over its full life cycle, being something like design–develop–manufacture–operate–decommission–disposal. In its life cycle, the asset has a productive phase ('operate'), in which it generates value by producing products or services. Elaborating on the life cycle view even further, Labuschagne and Brent claim that the life cycles of the products or services that the asset produces should also be considered. Figure 3.5 visualizes how these three life cycles, 'project life cycle', 'asset life cycle' and 'product life cycle', interact and relate to each other. Including sustainability considerations in projects therefore suggests that all three life cycles should be considered.

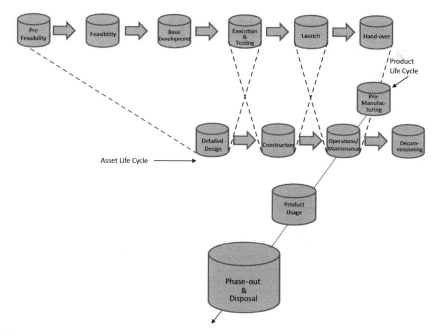

Figure 3.5   Interrelating life cycles
Source: Silvius et al., 2012, based on Labuschagne and Brent 2006

Because Labuschagne and Brent include the result of the project, the asset, in their framework, it is sensitive to the context of the project. Their studies regarded the manufacturing sector in which projects generally realize assets that produce products. In other contexts, the result of a project may not be an asset, but an organizational change or a new policy. The general insight gained from their work, however, may be that integrating sustainability in projects should not be limited to just the project management processes. It suggests that the supply chain of the project is also to be considered, including the life cycle of the resources used in the project, but also the life cycle of whatever results the project realizes. Integrating the concept of sustainability in project management may therefore very well stretch the systems boundaries of project management.

Another view on the scope of integrating sustainability in to projects can be found in the SustPM research project (Gareis et al. 2009 and 2011). This project focuses on integrating the concepts of sustainability specifically in project management processes and methods, and not the project management result or deliverable. This specific focus is motivated by the temporary character of projects, which causes the project management processes to be 'overlooked' in organizations, when striving for a more sustainable business. In the SustPM study, the concept of sustainability is detailed in six characteristics: economic-oriented, ecologic-oriented, social-oriented, short, mid, long-term oriented, local, regional, global-oriented and value-oriented. The project management processes are subsequently confronted with these six characteristics in order to develop additions to the project management standards and methodologies.

Conclusion 3: Project management standards fail to address sustainability

This conclusion was most explicitly reached by Mohamed Eid in the 2009 book *Sustainable Development & Project Management* (Eid 2009). Eid studied the integration of sustainable development in construction project management. Some conclusions from his study include:

- Project management is an efficient vehicle to introduce a more profound change, not only to the construction industry's practice, but more importantly to the industry's culture.
- Project management processes and knowledge fall short of committing to a sustainable approach.
- Mapping sustainable development on to project management processes and knowledge areas identifies opportunities for introducing sustainability guidelines in all project management processes.

Eid identified a number of 'leverage points' where sustainable development can connect into project management. These leverage points include the contribution to business strategy, the business justification, the procurement strategy, the readiness for service and the benefits evaluation of a project. The leverage points cover the whole life cycle of the project.

It should be mentioned, that 'help may be on its way' with regards to the integration of the concepts of sustainability into project management standards. For example, the SustPM project mentioned earlier already reported some results (Gareis et al. 2009 and 2011) and other initiatives to develop tools and insights to integrate sustainability aspects into project management also resulted in a number of publications.

Taylor elaborated on his earlier appeal to the project management profession (Association for Project Management 2006), by publishing *A Sustainability Checklist for Managers of Projects* (Taylor 2008) and a book on *Sustainability Interventions for Managers of Projects and Programmes* (Taylor 2011). Taylor takes on a very practical approach and both publications contain lists of potential considerations and interventions that can be used for understanding the sustainability aspects of projects. Although the checklists lack a systematic approach to the concepts of sustainability, they are meaningful tools for translating the somewhat abstract concepts of sustainability to the daily work of the project manager.

Taylor also participated in the 2010 IPMA Expert Seminar 'Survival and Sustainability as Challenges for Projects'. This seminar featured several papers and discussions on the integration of sustainability in projects and project management. The report of the seminar included a checklist of sustainability aspects of projects as developed by the international group of experts in the seminar and a mapping of these aspects on project management processes, roles and responsibilities of project members and project management competencies (Knoepfel 2010).

A more academic study into the operationalization of sustainability in projects was done by Iris Oehlman (Oehlman 2011). She developed the so-called 'Sustainable Footprint Methodology' to analyze and determine the relevant social, environmental and economical impacts of a project. This methodology confronts the life cycle of a project, consisting of three phases: project pre-phase, project execution and operation of the asset, with the three pillars of 'the triple bottom line': people, planet and profit (PPP). Each of the nine cells of the resulting framework is detailed in a set of sustainability indicators relevant to the respective sustainability perspective and the phase in the project life cycle.

A somewhat different approach is taken by Richard Maltzman and David Shirley in their book *Green Project Management* (2010). As the title of the book suggests, it focuses on the integration of environmental sustainability in project management. The book provides essential factual knowledge about environmental aspects and includes an extensive description of how project managers and sponsors can integrate these aspects into the different phases of a project. Maltzman and Shirley suggest that environmental aspects should be seen as aspects of quality and they introduce the term 'greenality' as the merger of 'green' aspects and the 'quality' of the project.

Conclusion 4: The integration of sustainability may change the project management profession

The conclusion of the 2010 IPMA Expert Seminar mentioned earlier was that the influence of project managers on the sustainability aspects of their projects at hand is substantial (Knoepfel 2010). This influence is regardless whether the project manager actually bears responsibility for these aspects of the project. Project managers have a strong influence because they are either responsible for certain aspects or they, in case they are not responsible, can influence the persons that are responsible.

This conclusion may actually change the nature of the project management profession. From a managerial role aimed at realizing delegated tasks, it may need to develop into a more advisory or leadership role with autonomous professional responsibilities and aimed at the right organizational changes.

The studies and conclusions summarized above illustrate the current state of knowledge on sustainability in project management. The current state of research on sustainability in projects and project management is mostly interpretive, giving meaning to how the concepts of sustainability could be interpreted in the context of projects, rather than prescriptive, prescribing how sustainability should be integrated into projects. Different authors pose different ideas and insights, containing many interesting suggestions about how project management should develop. However, most ideas and suggestions are of a rather conceptual nature and need elaboration to be of more practical value for the profession. The studies provide ingredients and provide questions, rather than answers. Some important questions are:

- Which system boundaries should be taken into account when considering sustainability in projects and project management?
- How can sustainability in projects and in project management be defined?
- How can the concepts of sustainability be made practically applicable to projects and project management?
- What are the implications of the integration of sustainability in projects and project management for project managers, projects and organizations?
- How can we measure and monitor the development towards more sustainable projects and project management?

The final sections of this chapter will explore the first two questions in order to develop a definition of sustainability in projects and project management. Based on this definition, Chapter 4 will explore the implications and application of this integration of sustainability on the level of the project management professional, the project management process and the project performing organization.

3.4   SCOPE OF SUSTAINABILITY IN PROJECTS AND PROJECT MANAGEMENT

The discussion of the different studies and publications on sustainability and project management in the last section (3.3) shows that the question of what sustainability means for projects and project management, cannot be answered without discussing the scope or system boundaries of project management. What should be considered in scope and what not?

Understandably, this question creates debate amongst both researchers and practitioners. For example Labuschagne and Brent (2006) include in their view on sustainability in project management the consideration of sustainability aspects of the project, the project's result or deliverables, and the effect that this result may have. Thereby suggesting that integrating sustainability in project management cannot be limited to the project delivery processes as such. The SustPM project, on the other hand, limits the consideration of sustainability aspects to the project management processes of a project. And next to these approaches, other definitions of the scope of considering sustainability in project management can also be defended.

The positioning of projects as temporary organizations that deliver changes to a permanent organization, as illustrated by Figure 3.3 provides a framework to illustrate the different approaches. Figure 3.6 illustrates five approaches we developed from literature and logical reasoning. We made an effort to provide logical names for the different approaches, but of course these are debatable. The different approaches are illustrated by indicating which processes are in scope for each approach (highlighted in black), on the overview provided in Figure 3.3.

When the concepts of sustainability are considered only on the level of the project management processes, for example, labour circumstances of project members or teleconferencing as alternative for travelling to meetings, we talk about 'Sustainability in project management processes'. This approach can be found, for example, in the earlier mentioned SustPM project.

The 'Sustainability in project delivery' approach also considers the sustainability aspects of the actual delivery of 'production' processes of the project. These processes can be of an informational nature, for example in software development projects, but also of a physical nature, for example in a project to produce gadgets or give-aways as part of a promotion campaign. The 'Sustainability in project management' approach also includes the resources and materials used in these project delivery processes as objects of consideration.

System boundaries of considering sustainability in project management

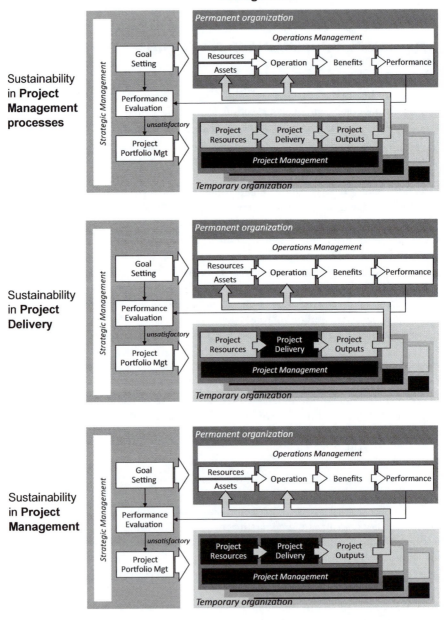

Figure 3.6   Different scopes of considering sustainability in project management

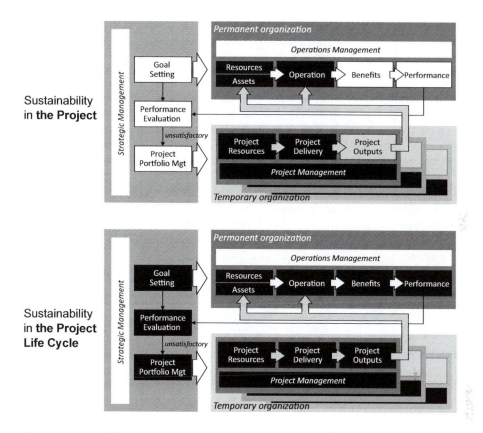

Figure 3.6   Different scopes of considering sustainability in project management (*concluded*)

In the 'Sustainability in the project' approach, the aspects of sustainability are also applied to the result or deliverable of the project. As stated earlier, this result can be any kind of change in organizations, policies, products, services, processes, assets and/or resources. The 'Sustainability in the project life cycle' approach adopts the views taken by Labuschagne and Brent and expands the consideration of sustainability aspects to the project, its results and its effects.

Figure 3.6 illustrates that different interpretations exist for the scope or system boundaries of considering sustainability in projects and project management. However, based on the 'Sustainability is about both short-term and long-term orientation' principle of sustainability, identified in section 2.5, it is inevitable to include the result of the project in the consideration of sustainability aspects. This identifies either the 'Sustainability in the project' or the 'Sustainability in the project life cycle' approach as most suitable.

3.5   DEFINITION OF SUSTAINABILITY IN PROJECTS AND PROJECT MANAGEMENT

Having defined the system boundaries of considering sustainability in project management, we can now elaborate on what the integration of all six principles of sustainability may imply for projects and project management.

Sustainability principle 1: Sustainability is about balancing or harmonizing social, environmental and economic interests

Corresponding with the TBL concept of sustainability, integrating sustainability in projects and project management requires the inclusion of 'People' and 'Planet' performance indicators in the management systems, formats and governance of projects (Silvius et al. 2012). In the current project management methodologies, the management of projects is dominated by the 'triple-constraint' variables: time, cost and quality (Project Management Institute 2008). And although the success of projects is most often defined in a more holistic perspective (Thomas and Fernandéz 2007), this broader set of criteria does not reflect on the way projects are managed. The triple-constraint variables clearly put emphasis on the profit 'P'. The social and environmental aspects may be included as aspects of the quality of the result, but they are bound to get less attention.

Sustainability principle 2: Sustainability is about both short-term and long-term orientation

As argued above, the inclusion of the long-term orientation stretches the system boundaries of project management beyond what is typically considered as the domain of project management. In the integrated life cycles view demonstrated by Labuschagne and Brent, one could argue that the boundary between the temporary project organization and the permanent organization does not exist anymore. This conclusion, however, confuses managerial responsibility and scope of consideration. The temporary nature of any project organization implies that the managerial responsibility of the project manager is also temporary. This temporariness, however, does not imply a short-term orientation. Since benefits of the project deliverable most likely occur after the project has been completed, a longer-term orientation, longer than the project life cycle, ahould also be included in the scope of consideration of the project.

Sustainability principle 3: Sustainability is about local and global orientation

In an increasingly global business world, more and more projects also touch upon the geo-economic challenges faced by society. Part of the project team may be located in emerging economies like India or China. Suppliers may be from all over the world. Similar for customers benefitting from the project's deliverables.

It is clear that the globalizing business world would also include globalized projects and project management. The impact of this globalization trend effects how project teams are organized and managed. Topics like virtual organizations and differing cultures get increasing attention in publications and on seminars. However, the focus of these publications seems to be limited to the processes of performing the project. Within the project management community, the discussion about globalization aspects of the results or deliverables of the project still has to emerge.

Sustainability principle 4: Sustainability is about consuming income, not capital

This aspect of sustainability refers to the principle that resources are not exhausted more or quicker then they may be regenerated. In projects this principle may apply to the use of materials, energy, water and other resources, both in the process of delivering the project or as included in or part of the deliverables of the project. Since the temporary nature of projects can create an extraordinary pressure for project members, this principle may be explicitly applied to the HR in the project. Their performance in the projects should not have a negative effect on their future ability to perform.

Sustainability principle 5: Sustainability is about transparency and accountability

The principle of accountability is already clearly present in project management. Project managers account for their actions and decisions in regular progress reports. These reports typically include the 'triple-constraint' variables: time, cost and quality. Following the earlier mentioned TBL concept, however, integrating sustainability implies that project managers are also accountable for the ecological and social aspects of their projects. This logically implies that project progress reports also include environmental and social indicators.

The principle of transparency is less obvious in project management. The principle of transparency suggests that project managers communicate potentially relevant events, considerations and decisions to (potential) stakeholders. In projects, however, it is common practice that a project manager controls the information flows from the project team to stakeholders, most often with the goal to influence or manipulate the stakeholder's perception of the project. From a stakeholder management perspective, this practice of influencing perceptions is quite logical and rational, but with an expanding set of potential stakeholders, the integration of sustainability may require a more transparent communication with all (potential) stakeholders.

Sustainability principle 6: Sustainability is about personal values and ethics

As discussed earlier, our behaviour as professionals and consumers reflects the values and ethical considerations of our society and of ourselves. Projects and project managers are no strangers to this. Also projects have specific values, norms and rules (Gareis et al. 2009). These values are influenced by the project context, but also by the project manager. For project managers, professional ethics and values are written down in professional 'codes of conduct'. Members of the PMI receive the *PMI® Code of Ethics and Professional Conduct* and most IPMA member associations also have a code of conduct for their members. The contents of these codes most often relate to the relationship of the project manager with the project sponsor and the relationship of the project manager with the association that he or she is member of. Some codes, most prominently the PMI one, also mention a responsibility of the project manager towards the society in general. Article 2.2.1. of the *PMI® Code of Ethics and Professional Conduct* states that 'We make decisions and take actions based on the best interest of society, public safety, and the environment.'. The statement clearly connects ethics and professional conduct with the concepts of sustainability.

Considering the perspective of projects as an instrument of change, the holistic nature of the sustainability principles and the conclusion on the appropriate scope to be considered, we developed the following definition of 'Sustainability in Projects and Project Management':

> *Sustainability in projects and project management is the development, delivery and management of project-organized change in policies, processes, resources, assets or organizations, with consideration of the six principles of sustainability, in the project, its results and its effects.*

3.6   A SUSTAINABILITY CHECKLIST FOR PROJECTS AND PROJECT MANAGEMENT

The reference in the definition above to the six principles of sustainability is powerful, but not yet very practical. The principles are by nature not a 'checklist' of sustainability aspects that can be checked or assessed. Most of the principles represent considerations or dilemmas that provide a different perspective on the project at hand. However, in order to integrate these principles into projects and project management, the 'object' of consideration, the project, needs to be specified in all its aspects.

A logical starting point for this specification can be found in the sustainability criteria and measurement systems mentioned in section 2.6. The 2010 IPMA Expert Seminar took the Global Reporting Initiative (GRI) sustainability reporting indicators as a starting point. Based on these GRI indicators, a checklist for considering sustainability in projects and project management was developed. Table 3.2 shows this checklist (Silvius 2010).

The checklist provides 11 categories of indicators with more specific indicators per category. These indicators were selected based upon their expected relevance

Table 3.2   A checklist for integrating sustainability in projects and project management

Economic Sustainability	Return on Investment	• Direct financial benefits/Net Present Value • Strategic value
	Business Agility	• Flexibility/optionality in the project • Increased business flexibility
Environmental Sustainability	Transport	• Local procurement/supplier selection • Digital communication • Travelling • Transport
	Energy	• Energy used • Emission/CO_2 from energy used
	Water	• Water usage • Recycling
	Waste	• Recycling • Disposal
	Materials and resources	• Reusability • Incorporated energy • Supplier selection
Social Sustainability	Labour Practices and Decent Work	• Employment • Labour/management relations • Health and safety • Training and education • Organizational learning
	Human Rights	• Non-discrimination • Diversity and equal opportunity • Freedom of association • Child labour • Forced and compulsory labour
	Society and Customers	• Community support • Public policy/Compliance • Customer health and safety • Products and services labelling • Market communication and advertising • Customer privacy
	Ethical behaviour	• Investment and procurement practices • Bribery and corruption • Anti-competition behaviour

Source: Silvius 2010

for projects in general. We suggest that a checklist like this is broken down further into specific types of projects or industries, for example information technology, construction, and so on. In comparison to the GRI, this checklist presents the indicators in a more concise manner and adds the 'business agility' indicator, which emphasizes the perspective of change, that is, characteristic for projects.

A checklist like this can be used by project managers and sponsors to review their existing project. Also the integration of the most important indicators in project progress reports would be a logical further step. A further elaboration on how the concepts of sustainability can be made practically applicable to project management follows in Chapter 4.

REFERENCES

Association for Project Management (2006), *APM supports sustainability outlooks*, http://www.apm.org.uk/page.asp?categoryID=4, Accessed 14 June 2010.

Barnes, M., (2002), A Long Term View of Project Management – its Past and its Likely Future, International Project Management Association, 16th World Congress, Berlin.

Crawford, L.H. and Pollack, J. (2004), Hard and soft projects: a framework for analysis, in *International Journal of Project Management*, 22 (8): 645–653.

Dombkins, D.H. (2006), *Competency Standard for Complex Project Managers*, College of Complex Project Managers And Defence Materiel Organisation, Commonwealth of Australia.

Dombkins, D.H. (2009), Redefining My Our Profession: The History and Future of Project Management, International Project Management Association, 22nd World Congress, Rome, Retrieved 12 August, 2011. Available at: http://www.pmforum.org.

Eid, M. (2009), *Sustainable Development & Project Management*, Lambert Academic Publishing, Cologne.

Foti, R. (2001), Forecasting the future of project management, in *PM Network*, 15 (10): 28–31.

Gareis, R. (2010), Changes of organizations by projects, in *International Journal of Project Management*, 28(4): 314–327.

Gareis, R., Huemann, M. and Martinuzzi, A. (2009), *Relating Sustainable Development and Project Management*, IRNOP IX, Berlin.

Gareis, R., Huemann, M. and Martinuzzi, A. (2011), What can project management learn from considering sustainability principles?, in *Project Perspectives*, Vol. XXXIII: 60–65.

Gorrino-Arriaga, J.P. and Eraso, J.C. (2000), Future Trends of Project Management, International Project Management Association, 15th World Congress, London.

Hartman, F.T. (2001), The Key to Enterprise Evolution – Future PM., PMI Seminars and Symposium Proceedings, Project Management Institute, Newtown Square, PA, USA.

Heerkens, G.R. (2001), How to Become the Successful Project Manager of the Future – Be Business Savvy! PMI Seminars and Symposium Proceedings, Project Management Institute, Newtown Square, PA, USA.

Ingason, H.T. and Jonasson, H.I. (2006), Exploring the Content of Project Management, International Project Management Association, 20th World Congress, Shanghai.

Jaafari, A. (1998), Project Managers of the Next Millenium: Do They Resemble Project Managers of Today?, International Project Management Association, 14th World Congress, Slovenia.

Kloppenborg, T.J. and Opfer, W.A. (2002), The current state of project management research: trends, interpretations, and predictions, in *Project Management Journal*, 33 (2): 5–18.

Knoepfel, H. (Ed.) (2010), *Survival and Sustainability as Challenges for Projects*, International Project Management Association, Zurich.

Labuschagne, C. and Brent, A.C. (2006), Social indicators for sustainable project and technology life cycle management in the process industry, in *International Journal of Life Cycle Assessment*, 11 (1): 3–15.

Maltzman, R. and Shirley, D. (2010), *Green Project Management*, CRC Press, Boca Raton, FL, USA.

McKinlay, M. (2008), Where is Project Management Running to…?, International Project Management Association, 22nd World Congress, Rome.

Oehlmann, I. (2011), *The Sustainable Footprint Methodology*, Lambert Academic Publishing, Cologne.

Project Management Institute (2008), *A Guide to Project Management Body of Knowledge (PMBOK® Guide)*, Fourth edition, Project Management Institute, Newtown Square, PA, USA.

Project Management Institute (2010), *Code of Ethics and Professional Conduct*, Project Management Institute, Newtown Square, PA, USA.

Rand, G.K. (2000), Critical chain: the theory of constraints applied to project management, in *International Journal of Project Management*, 18 (3): 173–177.

Russell, J. (2008), Corporate Social Responsibility: What it Means for the Project Manager, in *Proceedings of PMI Europe Congress*, Malta, Project Management Institute, Newtown Square, PA, USA.

Silvius, A.J.G. (2010), Workshop report Group 2, in Knoepfel, H. (Ed.), *Survival and Sustainability as Challenges for Projects*, International Project Management Association, Zurich.

Silvius, A.J.G., Brink, J. van der and Köhler, A. (2012), The impact of sustainability on Project Management, in *The Project as a Social System*, Henry Linger and Jill Owen (Eds.), pp. 183–200, Monash University Publishing, Victoria, ISBN:978-1-921867-04-0.

Taylor, T. (2008), A sustainability checklist for managers of projects. Retrieved on 28 December, 2010. Available at: http://www.pmforum.org.

Taylor, T. (2011), *Sustainability Interventions - for Managers of Projects and Programmes*, The Higher Education Academy – Centre for Education in the Built Environment, Salford.

Thomas, G. and Fernandéz, W. (2007), The Elusive Target of IT Project Success, International Research Workshop on IT Project Management (IRWITPM), Association of Information Systems, Special Interest Group for Information Technology Project Management.

Turner, R., Huemann, M., Anbari, F. and Bredillet, C. (2010), *Perspectives on Projects*, Routledge, Abingdon.

Woollett, J. (2000), Innovate or Die – the Future for Project Management, Prosperity through Partnership. World Project Management Week. Incorporating Project Management Global Conference, AIPM (CD-Rom).

World Commission on Environment and Development (1987), *Our Common Future*, Oxford University Press, Oxford.

Xia, W. and Lee, G. (2004), Grasping the complexity of IS development projects, in *Communations of the ACM*, 47 (5): 69–74.

CHAPTER 4

INCORPORATING SUSTAINABILITY IN PROJECT MANAGEMENT

GILBERT SILVIUS, RON SCHIPPER AND JASPER VAN DEN BRINK

The previous chapter demonstrated that the future of project management requires project managers to develop an increased business orientation in which sustainability is definitely a growing concern. By applying the six principles of sustainability to projects and project management, this orientation on business and sustainability can be developed into more specific impacts, guidelines and instruments. This chapter analyzes the practical integration of sustainability into projects and project management, on three levels:

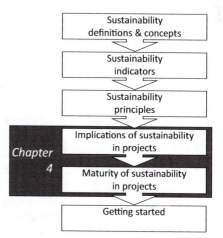

- What is the impact on the personal level: How will it affect project management competences and responsibilities?
- What is the impact on the project level: How can sustainability be integrated into the project management processes?
- What is the impact on organizational level: How do the impacts found on the personal level and the project level influence the way an organization performs or governs projects?

In this analysis of the impact of sustainability on different levels, we will make use some of of the most used standards for project management, such as *PMBOK® Guide*, PRINCE2®, ICB® VERSION 3.0, *PMCD Framework* and P3M3®.

Since the competences of the project manager are logically determined by the job the project manager has to do, the project management processes, we will first analyze the impact on the level of the project.

4.1   SUSTAINABILITY AT THE PROJECT LEVEL

This section discusses the integration of the concepts of sustainability at the project level. The first part of this section explores the impact of the aspects of sustainability on the processes of project management. The second part presents a particular maturity model for the incorporation of sustainability in projects and project management.

4.1.1   Sustainability in project management processes

The principles of sustainability and the sustainability checklist derived from these in section 3.6, are expected to have an impact on the way projects are performed and managed; for example in the identification of stakeholders or benefit areas, a logical starting point for the identification of this impact are the processes of project management. Two frequently used frameworks or standards for project management processes are the Project Management Institute's (PMI) *Project Management Body of Knowledge* (*PMBOK® Guide*) and the Office of Government Commerce's PRojects In Controlled Environments (PRINCE2®).

The *PMBOK® Guide* depicts the following project management process groups or clusters (Figure 4.1).

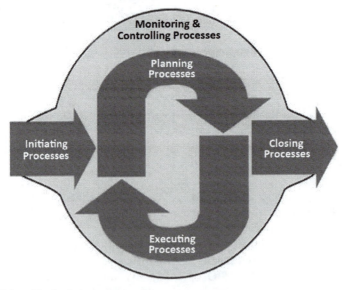

Figure 4.1   Project management process groups
Source: Project Management Institute, 2008

Although these process groups suggest a project life cycle, they do not necessarily represent the phasing of the project. Projects appear in many variations and 'there is no single way to define the ideal structure for a project' (Project Management Institute, 2008). The project management process groups may therefore depict a full project life cycle or the nature of the project management processes in a certain stage or phase of the project.

PRINCE2® bases its identification of project management processes on the project life cycle and identifies seven project management process groups: Starting Up a Project, Initiating a Project, Directing a Project, Controlling a Stage, Managing Project Delivery, Managing Stage Boundaries and Closing a Project (Office of the Government Commerce 2009).

When comparing these two models, the early and final process groups ('initiating' and 'closing') show great similarity. Also the controlling processes ('directing' and 'monitoring and controlling') show a certain similarity. For the middles stages of the project, the *PMBOK® Guide* shows a more iterative character than the PRINCE2® model. The PRINCE2® processes 'Managing Stage Boundaries' and 'Controlling a Stage', however, also imply iterative steps. Given the similarities between the two standards, we will use a generic project life cycle model as our baseline for analysing the impact of sustainability on project management processes.

In a study by Eid (2009), a forum of project management practitioners were asked about their assessment of the impact of sustainable development on project management processes. More specifically, for each project management process group (initiating – planning – executing – controlling – closing) the study asked their views on the area of integration of sustainability aspects. The questions asked were:

- To what extent do the scope and objectives of the project (more or less the *project content*) provide opportunities for integrating sustainability?
- To what extent do the actual processes of delivering and managing the project (the *project process*) provide opportunities for integrating sustainability?

Figure 4.2 shows the result of Eid's study.

From this study it can be concluded that the respondents see opportunities for the integration of sustainability in all process groups of project management. The area of integration of sustainability aspects, however, differs:

- The initiating and planning process groups provide opportunities for integrating the concepts of sustainability into the content of the project.
- The executing and controlling process groups provide opportunity for integrating the concepts of sustainability in the process of the project.

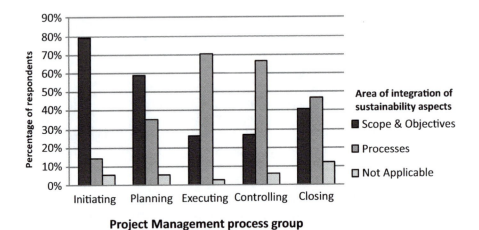

Project Management process group

Figure 4.2   The best areas to integrate sustainable development into project management

Source: Derived from Eid, 2009

This outcome is not unexpected, since the aspects of sustainability are best integrated from the origins of the process, system or asset that is defined as the intended project result. Again this conclusion points towards an involvement of the project manager already in the early and initiating stages of the project. In the executing and controlling process groups, the opportunities for considering sustainability shift to the execution of the actual project delivery and project management processes. For example, in waste handling, travel policies, job opportunities, labour conditions, and so on.

The closing processes of the project show a more diverse picture of the most opportune area to integrate sustainability aspects. According to Eid, the inclusiveness of the respondents may well suggest that the closing phase of a project offers the least appealing opportunity for integrating sustainability. However, we should point out the importance of the closing processes also for a more sustainable project result. The closing processes typically include handover to the permanent organization. The success of this handover and the acceptance of the project result are important aspects of a project's sustainability. Failed or non-accepted projects can hardly be considered sustainable, given the waste of resources, materials and energy they represent.

A more specific assessment of the impact of sustainability aspects and principles on the project management processes is presented in the following section. This section confronts the six 'principles' of sustainability, as derived in Chapter 2, with project management processes of a generic project life cycle. The terminology is based on the *PMBOK® Guide*.

Sustainability principle: Sustainability is about balancing or harmonizing social, environmental and economic interests

This principle provides additional perspectives on the content and process of the project. It logically influences both the content and the process of the project. Hence, it may be expected that this principle has a high impact on almost all project management processes. For example balancing social, environmental and economic interests will influence the project initiating processes in terms of the intended project result and the identification of stakeholders. Also in the project planning stage, providing additional perspectives will logically affect processes like defining the requirements, scope, activities, quality and risks of the project. The checklist of sustainability aspects, developed in section 3.6, provides guidance on these perspectives. Taking into account social and environmental criteria will logically also affect the organization of the project and the criteria for materials and supplier selection.

All considerations that apply to the project planning stage of the project will also affect the project execution phase. For example, additional perspectives on quality will logically also influence the quality assurance process. And applying social criteria like equal opportunity, also apply in the project execution stage, when organizing and managing the project team.

The project monitoring and control processes may be less strongly affected by the inclusion of social and environmental aspects. Of course the project content that is being monitored is influenced by the principal, but the monitoring and controlling processes itself not that much. One exception to this is the reporting process. The project's progress reports will definitely be influenced by the principle, for example by including reporting items that refer to the social and environmental aspects. Inclusion of these aspects will also influence the process of data gathering and preparing the report.

Taking into account social and environmental criteria is not expected to influence the project closing processes.

Sustainability principle: Sustainability is about both short-term and long-term orientation

This principle also provides an additional perspective on the content and the process of the project. For example, including the long-term future as a consideration may influence the selection of materials used in the project, favouring more eco-friendly materials. In this regard, the principle seemingly overlaps with the balancing social, environmental and economic interests principle, but it adds a time scale to these interests. Adding this time scale logically impacts almost all project management processes in the project initiating and project planning stages, such as defining intended outcomes, requirements, scopes, stakeholders, activities, qualities and risks of the project.

In the project execution stage, adding a long-term perspective could include developing team members beyond the needs of the project, but thereby building the organization's strength to perform projects. This would influence the processes of managing and developing the project team. Another consideration that is emphasized by the short-term and long-term process is the acceptance of the change in products, services, systems, processes, resources or procedures that the project includes. It is impossible to narrow this down to one process in the project. The acceptance of the project result is influenced by the way all project management processes are executed.

Following the reasoning stated earlier, the project monitoring and control processes may be less strongly affected by the inclusion of the short-term and long-term perspective. In the project closing stage, the short-term and long-term principle may be expected to emphasize the handover of the project result to the permanent organization and the acceptance of the organizational change that the project created.

Sustainability principle: Sustainability is about local and global orientation

In a similar way to the first two principles, this principle also provides a specific perspective on the content and the process of the project. For example, the global aspects of a project may include the labour conditions of organizations in the supply chain that may reside in low-cost countries like India, China or Africa. Local aspects may include engaging with stakeholders in the local community about the effect of the project (for example, noise or traffic related to the project) on their living environment. Since considering the local and global perspective can impact again on both the content and the process of the project, almost all project management processes in the project initiating, planning and execution stages may be affected by this.

The project monitoring and control processes may be less strongly affected by the inclusion of this perspective, as may the closing stage of the project.

Sustainability principle: Sustainability is about consuming income, not capital

This principle of sustainability provides guidance for the materials and resources in the project, for example by selecting eco-friendly materials to be used in the project. The impact of this principle therefore concerns the content of the project, thereby logically impacting the processes of the project planning stage.

However the principle also applies to the processes of the project execution stage, for example by caring for the wellbeing of the project team members or other stakeholders of the project. The result-oriented nature of projects, combined with often limited resources, may create high work pressure for the team members or suppliers involved. If this pressure leads to fall-out of team members, this should also be considered as 'consuming capital', since it affects a person's future ability to perform. The impact of this principle is therefore also relevant for the processes of the project execution stage.

The impact of this principle on the project monitoring and controlling and the project closing phase may be less obvious.

Sustainability principle: Sustainability is about transparency and accountability

The principle of transparency and accountability does not provide a new or different perspective on the project's content or process, but concerns the openness and pro-activeness in information and communication towards stakeholders (including team and sponsor en thei communication) and the general public. This principle therefore logically applies to all project management processes. It concerns 'how' processes are done, but also 'what' is done in terms of communicating and engaging with stakeholders.

Sustainability principle: Sustainability is about personal values and ethics

Following a similar reasoning as the last principle above, the principle of values and ethics also logically applies to all project management processes and stages. It is not that much about 'what' is done, but most of all about 'how' this is done.

Table 4.1 summarizes the impact of the sustainability principles analyzed above. The expected impact of the principles on stages is indicated as 'High', 'Moderate' and 'Low'.

Based on the analysis above, the following 'areas of impact' can be concluded:

4.1.1.1 Project context
Project management processes should address questions such as: *How do the principles and aspects of sustainability influence the societal and organizational context of the project?* And: *How is this influence relevant or translated to the project?*

4.1.1.2 Stakeholders
The principles of sustainability, more specifically the principles 'balancing or harmonizing social, environmental and economic interests', 'both short term and long term' and 'both local and global', will likely increase the number of stakeholders of the project. Typical 'sustainability stakeholders' may be environmental protection pressure groups, human rights groups and nongovernmental organizations (NGOs).

Table 4.1   Expected effect of sustainability principles on project management processes

	Sustainability principles					
	Harmonizing social, environmental and economical interests	Both short term and long term	Local and global	Consuming income not capital	Transparency and accountability	Personal values and ethics
Project Initiating	High impact	High impact	High impact	Low impact	High impact	High impact
Project Planning	High impact	High impact	High impact	High impact	High impact	High impact
Project Executing	High impact	High impact	High impact	High impact	High impact	High impact
Project Monitoring and Controlling	Moderate impact	Low impact	Low impact	Low impact	High impact	High impact
Project Closing	Low impact	High impact	Low impact	Low impact	Moderate impact	Moderate impact

4.1.1.3 Project content

Integrating the principles of sustainability will influence the definition of the result, objective, conditions and success factors of the project, for example the inclusion of environmental or social aspects in the project's objective and intended result.

4.1.1.4 Business case

The influence of the principles of sustainability on the project content will also need to be reflected in the project justification. The business case of the project may need to be expanded to include also non-financial factors that refer to for example social or environmental aspects.

4.1.1.5 Project success

Related to the project justification in the business case, it should be expected that the principles of sustainability are also reflected in the definition or perception of success of the project.

4.1.1.6 Materials and procurement

The processes concerned with materials and procurement also provide a logical opportunity to integrate aspects of sustainability, for example non-bribery and ethical behaviour in the selection of suppliers.

4.1.1.7 Project reporting

Since the project progress reports logically follow the definition of scope, objective, critical success factors, business case, and so on from the project initiating and planning processes, the project reporting processes will also be influenced by the inclusion of sustainability aspects.

4.1.1.8 Risk management

With the inclusion of environmental and social aspects in the project's objective, scope and or conditions, logically the assessment of potential risks will also need to evolve.

4.1.1.9 Project team

Another area of impact of sustainability is the project organization and management of the project team. The social aspects of sustainability, such as equal opportunity and personal development, can especially be put to practice in the management of the project team.

4.1.1.10 Organizational learning

A final area of impact of sustainability is the degree to which the organization learns from the project. Sustainability also suggests minimizing waste. Organizations should therefore learn from their projects in order to not 'waste' energy, resources and materials on their mistakes in projects.

Now that we have assessed the potential project management processes and related areas on which the aspects of sustainability may have a logical impact, a next step is to understand how this impact is covered in the current project management standards. Table 4.2 shows an overview of the areas of impact, how these are included in *PMBOK® Guide* and PRINCE2® and how these impacts are addresses in 'true' sustainable project management.

Reflecting on the analysis above, we can conclude that on the logical areas of impact, the standards of project management processes (*PMBOK® Guide* and PRINCE2®) refer mostly implicit to sustainability considerations. More explicit identification of especially environmental and social aspects is lacking. Furthermore, the *PMBOK® Guide* and PRINCE2® standards of project management place emphasis on the processes of project management. The content of the project (objective, intended result, deliverable) are mostly considered as given. Integrating the concepts of sustainability, however, suggests that not just the process of delivering the project is considered, but also the content of the project itself. The integration of sustainability in projects and project management therefore requires a different approach. A suggestion for a more content-oriented approach is presented in the maturity model (based on Silvius and Schipper 2010) in the next section.

Table 4.2   Overview of the impact of sustainability on the project management processes

Area of impact	*PMBOK® Guide*	PRINCE2®	True Sustainable PM
Project context	Section 1.8, Enterprise Environmental Factors, mentions the organization's human resources and marketplace conditions as 'internal or external environmental factors that surround or influence a project's success'. But the section fails to more explicitly identify potential social or environmental interest resulting from sustainability policies as factors of influence.	PRINCE2® addresses the project context in several processes during the start-up and initiation stages of the project. The business case and intended project result may logically be linked to organizational goals and strategy. No mention is made of a larger societal context of the project.	The context of the project is addressed in relationship to the organization's strategy, but also in relationship to society as a whole.
Stakeholders	In Section 2.3, Stakeholders, or the definition of stakeholders in the Glossary, any reference to typical sustainability stakeholders as environmental protection pressure groups, human rights groups or NGOs are lacking. In fact, Chapter 10, Project Communications Management, also fails to recognize these potential stakeholders when it discusses stakeholder communication.	The identification of stakeholders is mentioned in different processes of the initial stages of the project. Also the communication to stakeholders is addressed explicitly. There is no explicit recognition of potential stakeholders representing the environmental and/or social aspects of the project.	In the identification of potential stakeholders, explicit notion is made of potential stakeholders representing the environmental and/or social aspects of the project. Communication with stakeholders includes proactive engagement with potential stakeholders.
Project content	In the introduction of Chapter 3 Project Management Processes, the *PMBOK® Guide* mentions a few criteria for a successful project. Here it is mentioned that the project manager should be able to 'balance the competing demands of scope, time, cost, quality, resources and risk'. In this section the *PMBOK® Guide* fails to recognize the social and environmental aspects as relevant factors in project success.	PRINCE2® mentions six project performance variables. These variables do not mention sustainability aspects explicitly, but they may be included in performance variables 'quality' and 'benefits'.	The content, intended result and success criteria are based on a holistic view of the project, including sustainability perspectives as 'economical, environmental and social', 'short term and long term' and 'local and global'.

Table 4.2   Overview of the impact of sustainability on the project management processes *continued*

Business case	Section 4.1.1., Develop Project Charter, mentions 'Ecological impacts' and 'Social needs' as potential benefits of a project when it discussed the business case.	The business case has a central role in PRINCE2®. In all stages of the project, specific processes are identified to define or update the business case. In the business case, benefits are addressed in general, without specifically addressing potential social or environmental benefits.	The business case addresses the 'triple bottom line' of economic, social and environmental benefits. Investment evaluation is done based on a multi-criteria approach of both quantitative and qualitative criteria.
Project success	As stated earlier, the *PMBOK® Guide* mentions compliance with the project's requirements and objectives and specifically the demands of scope, time, cost, quality, resources and risk as aspects of the project's success. No mention is made of social or environmental aspects, unless included in the project's requirements or objectives.	Also, PRINCE2® does not mention sustainability aspects as aspects of project success. In the six project performance variables, the requirements of the project sponsor play a central role.	The definition and perception of project success take into account the 'triple bottom line' of economic, social and environmental benefits as laid out in the business case, both in the short term as in the long term. This implies that the success of the project is assessed based on the life cycle of the project and its result.
Materials and procurement	Processes related to the selection of materials and procurement can be found in different sections of the *PMBOK® Guide*. For example section 3.4.20 Plan procurements, section 3.5.8. Conduct procurements, Chapter 12 Project Procurement Management. None of these sections include any references to sustainability aspects in for example the selection of suppliers or the selection of materials.	Materials and procurement are implicitly included in work-packages. No reference is made to the selection of materials and suppliers, based on sustainability criteria.	In the selection of materials and suppliers for the project, these decisions are also based on environmental and social considerations.
Project reporting	Project reporting processes can be found in the *PMBOK® Guide* in section 3.6.8. Report Performance and section 10.5 Report Performance. In these sections, project reporting focuses on progress and changes in the areas scope, schedule, cost and quality of the project. Reporting on sustainability aspects is not explicitly addressed, nor is the principle of transparency.	The 'Report highlights' process as part of 'Controlling a stage' reports the progress of the project in terms of the work packages, issues and changes. Reporting on sustainability aspects is not addressed, nor is the principle of sustainability.	Project reporting is pro-active and transparent. Project progress is reported on different aspects of the project, including environmental and social aspects.

Table 4.2   Overview of the impact of sustainability on the project management processes *concluded*

Risk management	Chapter 11, Project Risk Management, of the *PMBOK® Guide*, does mention a process and several techniques to identify risks. However, these techniques do not mention the possibility of environmental and/or social risks.	Also Risk, as one of the central themes in PRINCE2®, is addressed in many processes throughout the project life cycle. However, no explicit reference is made to environmental and/or social risks.	The risk identification and risk management processes include the identification and management of environmental and/or social risks
Project team	Chapter 9 of the *PMBOK® Guide*, Project Human Resource Management, shows little consideration of social sustainability aspects such as life–work balance, equal opportunity, part time job opportunities, and so on. Section 9.2.2., however, pays attention to 'virtual teams' and links this to team members working from home offices, potentially with mobility limitations or disabilities. Also the personal development of team members is addressed. The objective for this development, however, is the performance of the project team, without considering the effectiveness of team members in their professional life after the project.	PRINCE2® pays ample attention to the management and development of the project team. It does mention 'Design and appoint the project management team', but in later stages no reference is made.	The management and development of project team members is aimed at preparing them for their role in the project and keeping them fit for this role. But it also considers the effectiveness of team members in their personal and professional life after the project.
Organizational learning	Section 2.4.3 mentions 'Historical information and lessons learned' as part of the 'Corporate Knowledge Base' of the organization. However, this section lacks a more explicit reference to organizational learning or knowledge management in order to improve an organization's competence in doing projects.	The 'Lessons log' and the 'Lessons report' explicitly capture the lessons learned in a project. These lessons are explicitly addressed in the starting up stage of a project in the process 'Capture previous lessons'.	Lessons learned and previous experiences are explicitly captured during project execution and closing and are made to use in the initiation and start-up of new projects. This is done to improve an organization's competence in doing projects.

4.1.2   A maturity model for integrating sustainability in projects and project management

Maturity models are a practical way to 'translate' complex concepts into organizational capabilities and to raise awareness for potential development. They provide guidance for action plans and allow organizations to monitor their progress (Dinsmore 1998). Most maturity models are derived from the Software Engineering Institute's Capability Maturity Model (Carnegie Mellon Software Engineering Institute 2002) and consider maturity of processes. For example, project management maturity is in this context a measure of the organization's ability to perform project management and related processes in a controlled and optimized way.

As concluded in the previous section, the integration of sustainability in projects and project management, however, requires a more content-oriented approach. For that reason, a project's sustainability maturity is determined in terms of level of consideration. This approach is based on the development models of sustainability presented in section 2.4 that show that sustainability can be considered on different levels, from reactively preventing the harmful effects of doing business to proactively contributing to a more sustainable society by the products and services delivered.

In the maturity model, this development approach is applied to projects. Four levels of consideration are identified:

- A first logical level is the level of project resources. For example, using resources that provide the same functionality, but are less harmful for the environment, like using hybrid cars instead of normal fuelled cars. These actions can reduce the less sustainable effects of operating the organization, but do not take away the cause of non-sustainability.
- A second level of consideration is the business process in which the resources are used. In projects the business process is the project delivery process and the project management process. A more sustainable business process takes away the cause of non-sustainable effects instead of just limiting or compensating them. For example, optimizing a service management process in such a way that less travel is required.
- A third level of consideration is looking at the way the project delivers its value: the business model of the project. For example, considering off shoring of project activities or changing from a 'static', blueprint-like, project management approach to a more 'dynamic' or developing one in which the project and the intended result are developed along the way.
- A fourth and final level of consideration takes into account not only the project's resources, business processes or model, but also the result and benefits themselves. How does the project goal and intended result contribute to a more sustainable society?

Level of project resources

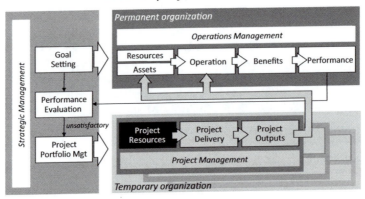

Level of project business processes

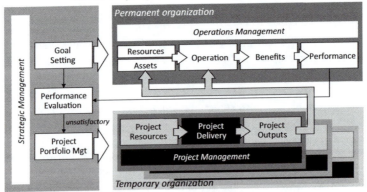

Level of project business model

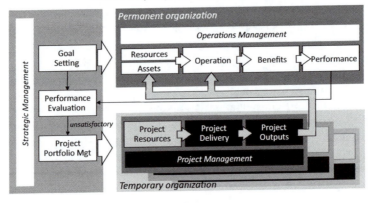

Figure 4.3   The levels of consideration of the maturity model

Level of project products/services

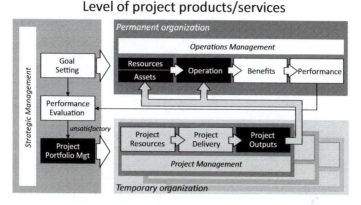

Figure 4.3   The levels of consideration of the maturity model *concluded*

Figure 4.3 plots these four maturity levels on the model we introduced in Chapter 3 to illustrate the scope of sustainability in projects and project management.

Based on the maturity model, an assessment instrument has been developed by the authors to assess the integration of sustainability concepts on the level of individual projects. This assessment uses a questionnaire consisting of four sections and 31 questions in total. The first three sections cover descriptive questions regarding the respondent, the project that is assessed and the organizational context of the project. The fourth section consists of the actual assessment questions. The model assesses the level (resources, business process, business model, products/services) on which the different aspects of sustainability are considered in the project. The sustainability aspects are derived from the sustainability checklist presented earlier and are grouped in economic aspects, environmental aspects and social aspects. Presenting the project's maturity separately on these three pillars of sustainability is a deliberate choice in order to address differences in ambitions or values an organization may have.

Figure 4.4 shows the conceptual model of the assessment.

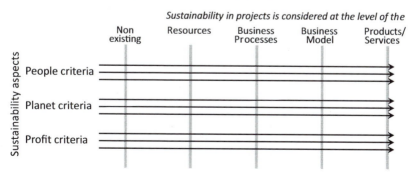

Figure 4.4   Conceptual model of the assessment

For each sustainability aspect an assessment of the current situation and the desired situation is asked. This provides guidance for improvement. Some example questions:

In which way does the project try to minimize its waste?

	Actual situation	Desired situation	
A.	[ ]	[ ]	No specific policies on this point.
B.	[ ]	[ ]	Waste in the project is separated in recyclable and non-recyclable and collected by the local waste handling companies.
C.	[ ]	[ ]	The project has policies (e.g. double sided printing) to minimize waste and waste in the project is separated.
D.	[ ]	[ ]	The project is designed to minimize waste and necessary waste is as much as possible recycled in the project itself.
E.	[ ]	[ ]	The project and the result it delivers are designed to minimize waste and necessary waste is as much as possible recycled in the project or result itself.

To what extent does the project apply policies or standards for diversity and equal opportunity that reflects the society it operates in?

	Actual situation	Desired situation	
A.	[ ]	[ ]	The project does not have any specific policies on diversity and equal opportunity, but complies with the standards and regulations of the organization it operates in.
B.	[ ]	[ ]	The project explicitly seeks diversity and complies with applicable standards and regulations on equal opportunity in terms of gender, race, religion, etc.
C.	[ ]	[ ]	The project actively (re) designs its work processes in a way (e.g. by designing part-time jobs) that diversity and equal opportunity are promoted and stimulated.
D.	[ ]	[ ]	The project actively (re) designs its work processes in a way (e.g. by designing part-time jobs) that diversity and equal opportunity are promoted and stimulated, and requires its suppliers to practice diversity practices and provide equal opportunity in terms of gender, race, religion, etc.
E.	[ ]	[ ]	The project's result is designed to improve diversity and equal opportunity in the society it operates in and this reflects in the way the project is executed and in its suppliers and users.

The full questionnaire is added as Annex C.

The result of the assessment is reported in a graphical way (Figure 4.5), showing both the actual levels and the desired levels of integration of the sustainability aspects.

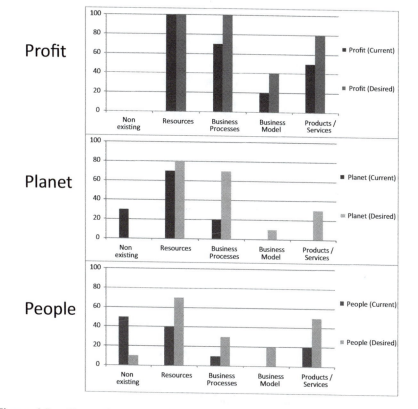

Figure 4.5   **Reporting format showing actual levels (dark colours) and desired levels (light colours) of integration of sustainability aspects**

Based on the maturity assessment, organizations can discuss their ambition levels (the desired situation) for the different perspectives and develop an action plan to bridge the gap between actual levels of maturity and desired levels and to monitor their progress on individual projects.

4.2   SUSTAINABILITY AT THE PROJECT MANAGER LEVEL

This section discusses the integration of the concepts of sustainability at the personal level, the project management professional. First, the responsibility of different roles for the sustainability aspects of projects will be discussed. After this, the standards on project management competences will be analyzed for competences that relate to the aspects and principles of sustainability. The final part of this section focuses on the impact of sustainability on the attitude and personal values of the project manager.

4.2.1   Responsibility for sustainability

Integrating sustainability at the personal level touches upon the question of responsibility. What is the responsibility of the project manager or the project team, for the sustainability aspects identified earlier? This question was also discussed in the 2010 International Project Management Association (IPMA) Expert Seminar (Knoepfel 2010). The participants of the meeting identified the different organizational roles in and around a project and discussed their responsibility for the different sustainability aspects from the checklist (section 3.6). The organizational roles that were identified were grouped into roles in the project team and roles in the context of the project.

The organizational roles in the context of the projects to be considered:

- the project sponsor, who is responsible for realizing the goal of the project in the context of his or her organizational responsibility;
- the portfolio manager, who has an initiating or coordinating role across several projects with the aim of realization the company's strategic goals;
- the programme manager, who is responsible for a set of interdependent projects and other activities, aimed at realizing a new company competence or strategic goal;
- the senior supplier and/or senior user, that are from their delivery and use perspective responsible for the content of the change.

Within the project team, the roles identified were:

- the project manager, who has often an important role in the organization and execution of the project and is therefore well positioned to create an environment in which sustainability aspects are integral to the way the project is executed;
- the designer, architect or technologist, who has the opportunity to consider sustainability in specifications of products/services, processes and supporting tools and techniques;
- the construction or realization manager, who is responsible for realizing the artefact or organizational change that the projects delivers.

Next to these roles in or around the project, the role of the project user, responsible for utilizing the project deliverable in the 'permanent' organization, was also identified.

For these different roles, their responsibilities for the criteria from the sustainability checklist was assessed. The responsibility was classified as 'is responsible for', 'can influence' or not applicable. Table 4.3 shows the result of this conceptual mapping (based on Silvius 2010).

Table 4.3   Mapping the responsibility for sustainability

		Project context				Project team			Operations
		Project sponsor	Portfolio manager	Program Manager	Senior user or Senior supplier	Project manager	Designer/ architect	Construction/ realisation manager	Project user
Economic Sustainability	Return on Investment	is responsible for	can influence	can influence	can influence	can influence	can influence	can influence	can influence
	Business Agility	can influence	can influence	can influence	can influence	is responsible for	is responsible for	can influence	
Environmental Sustainability	Transport	responsible or can influence	can influence	can influence	can influence	responsible or can influence	can influence		
	Energy	responsible or can influence	can influence	can influence	can influence	responsible or can influence	responsible or can influence	can influence	can influence
	Waste	can influence			can influence	responsible or can influence	is responsible for	is responsible for	can influence
	Materials and Resources	responsible or can influence			can influence	responsible or can influence	is responsible for	is responsible for	can influence
Social Sustainability	Labour Practices and Decent Work	responsible or can influence	can influence	can influence	can influence	is responsible for	can influence	is responsible for	is responsible for
	Human Rights	responsible or can influence	can influence	can influence	can influence	is responsible for	can influence	is responsible for	is responsible for
	Society and Customers	responsible or can influence	can influence	can influence	can influence	is responsible for	can influence	can influence	can influence
	Ethical Behaviour	responsible or can influence	can influence	can influence	can influence	is responsible for	can influence	is responsible for	is responsible for

Source: Based on Silvius 2010

From this mapping of responsibilities, some conclusions can be drawn:

- The project manager and his team have a clear role in realizing a more sustainable project, project results and therefore a changing organization and society. Either they have the responsibility for the different aspects of sustainability or they can influence them.
- The responsibility for sustainability in and of the project is mostly divided between project sponsor and project manager, with the exact arrangement depending on the content and context of the project. Although many aspects of the project may logically be determined by the project sponsors, or other roles in the context of the project, the question can be raised whether the responsibility for sustainability in projects and project management should not be considered a shared responsibility of project sponsor and project manager together? Or even of all major stakeholders? Either way,

the project manager is well positioned to have a strong influence on the sustainability of the project and its project management. And given the importance of projects for creating a more sustainable future, it can be questioned whether the project management profession can ignore a certain responsibility to apply this influence for a more sustainable development of their organization.

4.2.2   Sustainability competences

Taking a responsibility for sustainability requires adequate competences. Sustainability is a complex and holistic concept and it is advisable to understand which competences integrating sustainability in projects and project management, would be required of a project manager. This section will analyze the most used project management competence standards and assess how the principles of sustainability are reflected in these standards, or how they can be incorporated.

Standards for project management competences, for example, scheduling, risk management, quality management, and so on, first emerged as part of the standards of the International Council on Systems Engineering (INCOSE). However, the first integrated standards for project management competences did not appear until the mid 1990s (for example, Australian Institute of Project Management (AIPM), IPMA). In 1997, IPMA launched the first version of its 'International Competence Baseline' (ICB®). An improved second version followed in 1999 and a third one, the ICB® Version 3.0, in 2006. In the development of the ICB®, competences addressing the behavioural and change aspects of projects and the context of projects received increasing attention.

Today, the ICB® Version 3.0 is one of the most widely used standards for project management competences. It is frequently used by organizations as a framework for assessing and developing project managers. The ICB® Version 3.0 breaks project management competence down into 46 competences in the following categories:

- technical competences (20 competences) which cover the project management processes;
- behavioural competences (15 competences) which deal with the personal skills of the project manager and their relationships with stakeholders of the project;
- contextual competences (11 competences) which cover the interaction of the project with its context (projects, programmes, portfolios and the permanent organization).

These competences are illustrated in the 'eye of competence' (Figure 4.6) and are briefly described in Annex D.

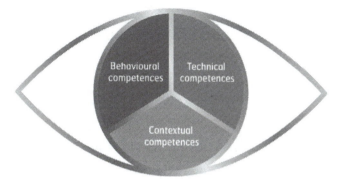

Figure 4.6   An overview of project management competences according to the ICB® Version 3.0
Source: International Project Management Association 2006

The ICB® Version 3.0 has little mention of the word 'sustainable': competence 2.04 Assertiveness talks about 'sustainable relationships to the interested parties' (page 94) and competence 3.09 Health, Safety, Security, Environment talks about 'security and sustainability' (page 32). The content of sustainability, however, is addressed under the key word of 'project context'. This starts in competence 1.3 Project Requirements & Objective, where the conformity to the context conditions is required in addition to achieving the project objectives. The context is later mentioned in several of the contextual competences: 3.05 Permanent Organization; 3.06 Business; 3.07 Systems, Products & Technology; 3.08 Personnel Management; 3.09 Health, Safety, Security, Environment. In 3.07 Systems, Products & Technology and 3.09 Health, Safety, Security, Environment the subjects within sustainability (for example, the systems life cycle management) are well addressed. Also, the responsibility (permanent organizations) and some processes (for example, internal and external audits) and tools (Environmental Impact Study) are mentioned. Also the reference to ethics in competence 2.15 implies that at least some social aspects are taken into consideration. Other references to the social aspects of sustainability can be found in element 3.08 Personnel Management and 2.14 Values Appreciation.

Another frequently used competence framework is the Project Management Competence Development (PMCD) Framework from PMI (Annex E). The *PMCD Framework* identifies three 'dimensions' of competence (Project Management Institute 2007):

- Knowledge: this refers to what the project manager knows about project management.
- Performance: this refers to what the project manager is able to do or accomplish while applying their project management knowledge.

- Personal: this refers to how the project manager behaves when performing the project or related activity.

The *PMCD Framework* describes the generic competences needed in most projects, most organizations and most industries. In some industries there may be specific competences needed, for example specific domain knowledge or knowledge of regulatory and legal requirements. Also, specific organizational knowledge may be required, for example about policies, procedures, internal organization or culture. Overall the *PMCD Framework* therefore identifies five 'units' of competences (Figure 4.7).

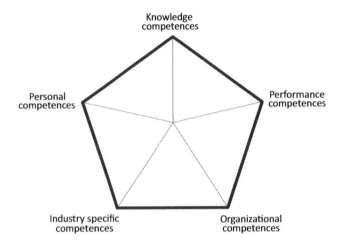

Figure 4.7   Overview of the *PMCD Framework*
Source: Project Management Institute 2007, adapted

The *PMCD Framework* actually mentions 'Knowledge' as a separate unit of competence, but also suggests that knowledge is logically included in all four other units.

The *PMCD Framework* has no mention of the words sustainability or social responsibility. Nevertheless some competences relate to sustainability aspects, for example, competence 1.1 'project aligned with organizational objectives and customer needs' and 1.2 'preliminary scope statements reflects stakeholder needs and expectations'. In these competences the project aligns with the goals and strategy of the organization and these may very well include goals on sustainability. With this alignment, sustainability goals may logically be part of the project goals. Another example is competence 6.1 'actively listens, understands and responds to stakeholders'. Application of this competence assures the active stakeholder engagement that is also suggested by the ISO 26000 guideline. By engaging with stakeholders, project managers actively discuss potential sustainability needs or wishes from stakeholders.

Another indirect link to sustainability is included in competence 9.1 'takes a holistic view of project'. Integrating the concepts of sustainability into a project needs a holistic view. Competence 11.2 'operates with integrity' hints at personal values and perhaps even ethics, which were identified as one of the principles of sustainability.

More links to sustainability can be found in the *PMI® Code of Ethics and Professional Conduct (*Project Management Institute 2010). The new code addresses the expanding global involvement in the project management profession. This code addresses the increased awareness of business ethics as well as the differentiation of ethical values in different cultures. The revised code of conduct addresses four values and further separates them into aspirational and mandatory components. The four areas are: Responsibility, Respect, Fairness, Honesty. The new code is meant to provide a single framework for PMI members and from 2011 on, sustainability will be given a more prominent place in the project manager assessment and certification process. The Professional and Social Responsibility content area will be tested in every domain rather than as a separate domain on the examination. The recognition obtained is that professional and social responsibility is integrated into all of the work of project management. The *PMI® Code of Ethics and Professional Conduct* should therefore be viewed as now integrated into the day-to-day role of a project manager, emphasizing its importance in each phase of the project life cycle. The conclusion is that there is a turning towards sustainability as part of projects and project management; but little or no concrete approach is given.

When both standards are studied, an overlap between the ICB® Version 3.0 and the *PMCD Framework* can be discovered at a conceptual level. The rudimentary analyses was performed for the purpose of choosing one of the two competence descriptions and the result is shown in Figure 4.8 (please note that the *PMCD Framework* competence unit 'knowledge' is treated as implicit in the other *PMCD Framework* units of competences).

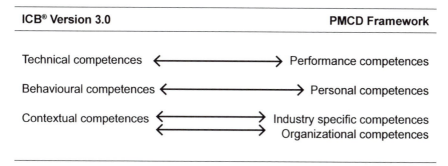

Figure 4.8 **Conceptual comparison of the ICB® Version 3.0 and the *PMCD Framework***

From both standards and their comparison the following observations can be made:

- Both frameworks consider a competence as the combination of knowledge, skills and personal attitudes required to be successful in a certain function or to complete a certain task or goal.
- In both standards, these three components (knowledge, skills and attitudes) are interwoven: the knowledge of the project management process, the personal skills required for applying this knowledge and the behaviour while applying.
- Both frameworks recognize internal aspects of managing the project (technical competences/performance competences), personal competences of the project manager (behavioural competences/personal competences) and external aspects (contextual competences/industry specific and organizational competences).

Since the two frameworks show a strong resemblance in their identification of relevant categories of project management competences, we will adopt the ICB® Version 3.0 structure of technical–behavioural–contextual competences for the further analysis of the integration of sustainability principles in the competences.

In this further analysis, the three categories of project management competencies are confronted with the six principles of sustainability. The principles will be used to determine the strength of, or influence of, each competence on realizing sustainability in the project result or process. The following section presents the outcome of this analysis.

4.2.2.1 Sustainability in the technical competences

Section 4.1.1 analyzed the impact of sustainability on the project management processes. It showed that especially the principles 'Harmonizing social, environmental and economic interests', 'Both short-term and long-term orientation' and 'Local and global orientation' provide a logical extension of the aspects and actors to be considered in the project. This extension will also reflect in the technical competences required for managing these processes, like risk and opportunity, interested parties, scope and deliverables, quality, control and reports.

The principles 'Consuming income, not capital' and 'Personal values and ethics' can be understood to apply to materials applied in the project, but also to the project organization (for example, not 'exploiting' team members to an unacceptable level), suppliers and sub-contractors (for example, 'fair' purchasing) and the purchasing process (for example, non bribery). Logically this will reflect in competences like project organization, resource management, procurement and contract, and so on.

The principle 'Transparency and accountability' logically relates to competences that involve the information and communication within and around the project, like control and reports, communication, cost and finance, information and documentation.

From this analysis, it should be concluded that the impact of sustainability on the technical competences of project management is substantial. On the one hand this conclusion may be surprising, because sustainability seems more of a contextual influence, but on the other hand it could be expected that the principles of sustainability may affect all aspects of projects and project management.

4.2.2.2 Sustainability in the behavioural competences
The ICB® Version 3.0 already includes values appreciation and ethics as behavioural competences, which makes the connection to the principle 'Personal values and ethics' very obvious. But values and ethics should also be expected to influence other competences, like negotiation, conflict and crisis, leadership, reliability, and so on. The principle 'Transparency and accountability' may logically influence the competence of openness since it may require a very 'open' personality to be able to be open, transparent and accountable to stakeholders.

The sustainability principle 'Consuming income, not capital', also relates to the behavioural competences of the project manager. It implies that he or she has an eye for the pressure put upon team members and manages the team in a 'sustainable' way (see section 4.2.4.).

Given the more 'personal' nature of the other behavioural competences, the other sustainability principles seem to have a more limited impact.

Based on this analysis, we classify the impact of sustainability on the behavioural competences as substantial, but mostly reflected in *how* the project is done and managed and not so much in *what* is done.

4.2.2.3 Sustainability in the contextual competences
With the competence health, security, safety and environment, the ICB® Version 3.0 includes a competence that clearly relates to the 'people, planet, profit' (PPP) concept of sustainability. But the principles 'Harmonizing social, environmental and economic interests', 'Both short-term and long-term orientation' and 'Local and global orientation' should be expected to influence more competences. For example: systems, products and technology, business, legal, permanent organization, portfolio orientation, and so on.

Following the arguments explained earlier, the principles 'Consuming income, not capital' and 'Personal values and ethics' can be understood to apply to the competence personnel management.

Based on this analysis, we classify the impact of sustainability on the contextual competences as rather substantial.

Summarizing the analysis and conclusion above, Table 4.4 shows the impact of sustainability on the different project management competence categories.

Table 4.4   Impact of sustainability on project management competences

	Sustainability principles					
	Harmonizing social, environmental and economical interests	Both short term and long term	Local and global	Consuming income not capital	Transparency and accountability	Personal values and ethics
Technical competences	High impact	High impact	High impact	High impact	High impact	High impact
Behavioural competences				High impact	High impact	High impact
Contextual competences	High impact	High impact	High impact	Moderate impact		Moderate impact

This table shows that the content-oriented sustainability principles, like 'Harmonizing social, economic and environmental interests', 'Both short-term and long-term orientation', 'Local and global orientation' and 'Consuming income not capital', have their impact mainly on the technical and contextual competences. This makes sense, because the contextual and technical competences correspond with the `why´ and `how' questions of the project. The more moral-oriented principles, 'Transparency and accountability' and 'Personal values and ethics' have their impact mainly on the technical and behavioural competences of the project manager. Also, this makes sense because the technical competences cover managerial activities like organizing and reporting. And the behavioural competences refer to the person of the project manager and his/hers behaviour.

4.2.3   Missing competences?

Another question is whether there the integration of sustainability leads to new project management competences. Are certain competences missing in the ICB® Version 3.0?

4.2.3.1 Consulting skills
In section 4.2.1 we concluded that the project manager is well positioned to have a strong influence on the sustainability of the project and its result and that the project manager may, or should, apply this influence in order to make organizations more sustainable. This role of influencing the stakeholders of the project and

the members of the project team is a role that typically requires advising or consulting skills, more than managerial skills. ICB® Version 3.0 does not mention consulting skills explicitly as one of the project management competences. It does mention 'Consultation' as one of the behavioural skills, but this relates more to seeking consultation than to consult or advise others. It should be concluded that the integration of sustainability in projects and project management would require project managers to also develop competence in consulting and that this competence should be included in the project management competence standards.

4.2.3.2 Sustainability knowledge

On a more cognitive level, it should be expected that the project manager should also develop more knowledge about sustainability aspects. The ICB® Version 3.0 mentions as one of the competences 'Health, security, safety and environment'. This competence should be expected to have a broad coverage of sustainability aspects, but lacks a more explicit mention of the sustainability aspects listed in the checklist of section 3.6. These sustainability aspects will likely expand the set of stakeholders of the project. Typical 'sustainability stakeholders' may be environmental protection pressure groups, human rights groups, NGOs and so on. In order to perform the project successfully, the project manager needs to acquire the buy-in of the stakeholders. This would require that the project manager has conceptual and operational knowledge about the knowledge domains of the now extended stakeholders group. Examples of these knowledge domains may include life cycle costing, cradle to cradle, waste handling, opportunities for digital communication, energy use, criteria for decent work and recycling techniques.

4.2.3.3 Handling complexity

In section 3.1 we suspected that the ongoing technological progress and the need for a more balanced approach to economic, environmental and social development may cause a new spur in the development of the project management profession. Based on the analysis of the impact of sustainability on projects and project management we can now conclude that the complexity of projects and project management will most certainly increase. Integrating sustainability considerations provides new perspectives and a more holistic view on project outcomes, business cases, risks and stakeholders. It is the job of the project manager to take these new perspectives into account; a job that therefore increases in complexity. For project management competences this means that the project manager should develop competences like being able to work in uncertainty.

With regards to 'missing' competences, we should conclude that it makes sense for the future-proof project manager to develop adequate consulting skills, build expertise in the aspects that determine the sustainability impact of the project and handle complexity in and around projects. As well as technical advice on sustainability aspects, the project manager, being the specialist on change, could also advise on the change aspects of the project. And since influencing the content of the project may be best done in the early stages of the project, it would make sense for organizations, to involve the project manager already then.

4.2.4   Values and attitude

From the analysis of project management competences in the previous section, we concluded that the principles of sustainability are partly covered in the most used competence standards and we suggested some additions in terms of knowledge and skills. But competences consist of three elements: knowledge, skills and attitude. So what does integrating sustainability means for the attitude of the project manager?

'In order to change the way we DO things, we need to change the way we VIEW things', was the wisdom of Global Reporting Initiative (GRI) Deputy Director Nelmara Arbex, we mentioned earlier. The competence descriptions address attitude in terms of the behaviour a project manager demonstrates, but lack a vision on where this behaviour originates from. This source is the value system of the individual project manager. A value system is the image a person has of the world with convictions connected to them. Convictions include what is good or bad, important and not important. When life conditions change and there are reasons to think differently, then people also change. An example of a value system model is the model of Graves (Annex F). It consists of different value system levels, with each level implying a more complex system.

So viewing the world from a sustainable perspective which is more complex than the traditional profit view requires adopting or developing a new value system. The individual project manager can develop a new value system when he or she accepts a certain responsibility for sustainability, both as an individual and as a professional. Accepting responsibility changes behaviour. As an individual this change could include preferring more sustainable products and services or using voting power within the democratic process. As a project management professional this change may include influencing the stakeholders of the project as described in section 4.1.1. Adopting or developing new value systems, drives change within companies and society. This corresponds with the trends where a project manager's attitudes in fulfilling their role shifts from technical and result-oriented towards a more goal and context orientation.

4.2.5   How to manage the project team in a sustainable way

Discussions about sustainability vary between different industries. For the construction industry the environmental part of sustainability is of huge importance. For international trading companies, issues concerning the people in the developing world are of huge importance. But for many areas in the service industry the environmental impact of projects is not that big, neither is the impact on the living conditions in the developing world (for example, a reorganization of an IT section of a local healthcare institute), leading to the conclusion that sustainability isn't really an issue for them. These project managers are missing one crucial, but often neglected, part of sustainability. Sustainability isn't only about the global problems, it's local as well. In many projects the environment isn't the big issue, neither is poverty in the developing world countries really an issue. The major sustainability issue lies within the project, namely the well-being of the project members as well. When discussing environmental issues, poverty and shareholders' value, the humans who are working on the project are often forgotten. Even though the project members are, in most projects, the most crucial factor for project success. And in most western projects their health is only taken care of in a very limited way. This neglect of the project member is to a certain extent easy to understand because the main physical problems in the working environment have been solved during the last 100 years. Working hours have been reduced, excessively heavy work is done by machines, and noise and pollution in the workplace have been reduced. The workplace has become a much healthier place, at least in terms of physical health. But at the same time, the psychological pressure on workers in general and project members in particular is huge. In 2005 it was the second most reported health problem for workers in the EU and could be responsible for up to 50–60 per cent of all work days lost. An astonishing 20 billion euros was lost in 2005 in the EU as a result of work-related stress. In 2007 an Australian Insurance company reported annual costs for Australia of about 15 billion euros.

It is likely that projects in many cases are particularly places with a lot of stress. Projects are often dealing with new, unique problems where standard solutions don't fit. This inevitably leads to problems; as well as the innovative character of most projects and the tight deadlines. A limited amount of stress is no problem, but huge amounts of stress can and often will cause health problems.

In the last decade much attention has been paid to ways to reduce stress in employees. Examples are training in relaxation techniques, breathing techniques or assertiveness, as well as training aimed at more healthy thinking styles. Many of these techniques are well known and effective. In our opinion this isn't enough. Dealing with the mental health of project members isn't only about reducing negative effects of work, it should concentrate on increasing the positive effects of work. A thoroughly sustainable project should create a condition in which the project members and the project manager flourish. Flourishing is different from 'not dealing with too much stress'. A workplace where there is not too much stress is neutral, a flourishing workplace is a positive workplace in which people feel well and perform well. Various studies have shown that these kind of positive, flourishing workplaces and projects are not only enjoyable to work within but are profitable for the company as well. A famous example is the clean-up and closure of a nuclear weapons production facility in the United States. The clean-up was estimated to take a minimum of 70 years and cost 36 million dollars. Following a change of leadership style into a positive project management style the project was finished in ten years, costing less than half of the budget while the facility was 13 times cleaner than required (Cameron 2008).

During the last decade, the new scientific field of positive psychology has done a lot of research into ways to create a more positive collaboration between persons in order to make them happier and to create more sustainable workplaces. This research has led to four strategies that can be used to create these kinds of workplaces and project teams.

4.2.5.1   Manage positive goals
One important way project members can carry out their work positively and increase the chances of high performance is the definition of project goals linked to their personal goals (Prat and Ashforth 2003). A project should have meaning for the project members, a purpose more than earning money; having goals in life and especially meaningful goals leads to increased positive effect.

For the performance of the team members it is important that the goal the team is working on is significant to them and links to their purpose in life. A good project that stimulates a positive work atmosphere and attitude should be meaningful. This means that it relates to higher values. For example, the goal 'to develop a new administrative system' does not provide much meaning. The goal 'to create an administrative system that will improve the work quality of the people who use it' has far more resonance for the project members involved. In order to fully use the meaningfulness of these goals it is important to link these overall goals to the daily activities of the project members; this will ensure that not only the whole project is meaningful to them, but their specific job for the next three hours is purposeful as well.

A second characteristic of a good project goal is that it is a difficult but achievable goal. A job that's too easy does not stimulate to people to work hard or enjoy what they do. At the same time a job too difficult isn't too stimulating either - it is is just discouraging. People flourish with a bit of stress as long as it is not too much and within their own control (can I influence my tasks?).

4.2.5.2   Manage positive emotions
Positive emotions lead to a number of different positive effects such as better negotiation skills, improved conflict management skills, increased willingness to help other people, improved creativity and problem solving, more social behaviour and increased motivation, all behaviour and skills that can be very useful for a project member and a project manager.

Since positive emotions can be so powerful, investing in them can be very fruitful. Take the time to invest in positive emotions in teams, especially when cooperation is important or when negotiations, conflicts or creativity is on the agenda.

This requires the project manager to perform a role not usually recognized as part of their remit. A good project manager who wants their project members to flourish is deliberately influences the atmosphere in the project team. This should be an explicit function of a positive project manager: to influence the emotional state of his project members (Cerny 2009). At the same time the project manager should deliberately influence their own emotional state and stimulate their own positive emotions. This is not the same as fighting stress and negative emotions, which will not automatically lead to positive emotions.

Stimulation of positive emotions can be used during the whole project to encourage a broader perspective on problems and to increase cooperation between team members. This can be fruitful during stressful periods when people tend to narrow their perspective and therefore are often less capable of solving problems creatively. People tend to cope with stress more effectively when they experience and show positive emotions (Folkman 1997). However it is useful at any time because positive emotions stimulate a flourishing project team at any moment during the project process.

There are numerous strategies to stimulate positive emotions in a project team; having a good laugh together is the simplest one but that may be difficult to organize, unless you have a second career as a stand-up comedian.

Emotional contagion is a useful technique. Emotional contagion is the process by which group members copy the emotion that other team members display non-verbally (Ashkanasy and Ashton-James 2007). Being critical and angry during contact with project members will make things worse. Positive contagion involves showing positive emotions during your contacts. This may sometimes mean that you need to 'cheer yourself up' before you go into a conversation. The process of emotional contagion is especially strong when the leader of a group (most often the project manager) is showing a certain emotion.

Another way to influence the level of positive emotions in a project team is by paying explicit attention to positive events (Seligman et al. 2005) or experiences for which you are grateful (Emmons and Cullough 2003). You can do this individually by drawing attention to them at the beginning or the end of the day and thereby influencing your own positive emotions. You can also draw the project team's attention to them. Make discussing what went well during the project an explicit part of the project meetings – it will serve the overall level of positive feelings and we can often learn from what went well in order to improve processes.

4.2.5.3   *Manage positive relations*

Probably the most influential way to encourage a flourishing project team is to stimulate positive relations in the team. Losada and Heaphy (2004) show the positive effects of a conduciveworking environment in a study they conducted on the use of language in different teams. They studied communication in management teams and compared successful management teams with less successful management teams – looking at 360 degree feedback, client satisfaction and profitability. The best performing teams were the teams where the most 'positive' language was used (supportive, encouraging and helpful language with plenty of positive feedback and an appreciative attitude). The most successful teams had a ratio of 5.6:1 between positive and negative language statements. The lowest performing teams had a ratio of 0.36:1, which means that in those teams there where about three negative remarks compared to one positive remark.

Losada and Heaphy showed the beneficial effects of positive relationships. They did not however claim that a (project) manager should be totally, 100 per cent positive. There is the need for critical feedback to correct behaviour that is damaging to the overall performance of the team or behaviour that is negative for the climate in the team. So feedback on negative performances should still be given. A project manager should be aware of the damaging effects of too much negative communication and the encouraging effects of a lot of positive communication.

4.2.5.4   *Manage strengths*

A last lesson that can be learned from positive psychology research is the explicit use of personal strengths in projects (Clifton and Harter 2003). Using your own strength and the personal strengths of project members is far more rewarding and productive then focusing on weaknesses. Working on strengths will give the project member a chance to excel, working on weaknesses only a chance to become competent.

The positive effect of the strength-based approach works through two different mechanisms – positive results and positive motivation. It's a simple but often neglected rule in many projects. Too often project members are obliged to spend part of their time doing work that does not play to their strengths, just because 'it is part of the job' or because 'everyone has to do it'. When a project member is not good in administration, be sure to limit their administrative tasks to a minimum; for example by delegating this task to a project member who is structured and organized.

The main reason why using the personal strengths of people works is that it leads to positive results. When project members are spending most of their time working on tasks in which they perform well, they will be more productive compared to project members that spend a larger part of their time on tasks to which they are not particularly suited.

A second reason why it works is as a result of the positive boost to your motivation when people are using their strengths. Overall enjoyment in work increases; individuals are more willing to go that 'extra mile' when they applying their strengths. Of course both positive performances and positive motivation are mutually reinforcing.

For project managers this means that they should be well aware of their own strengths (Morris and Garrett 2010) and of the strengths of their project members. This can be done by formal tests such as the VIA Me! Character Strenghts Profile from the VIA Institute of Character or the Strength Finder from the Gallup group, or by interviewing and observing people.

However, knowing your own strengths and that of your project members is not sufficient. In order to fully utilise the strengths available in a project team you should aim to adapt the way you organize the workload in your teams. Project managers should build tasks around people instead of building people around tasks. This leads to a new perspective on the division of work and the way the work breakdown structure (WBS) is being translated into tasks for project members. Thus the WBS is no longer the starting point for planning, but rather the strengths and availability of individual team members. Work in the project should be organized around the strengths of the individual project members and the project managers, as far as this is possible.

The big challenge for many project managers who want to work in a sustainable way is right in front of them in the management of their day-to-day business of their own project members. Using simple interventions, project members can seriously improve both the health and the productivity of their project teams. Of course these interventions are not limited to the project team. They can easily be applied in contacts with the sponsor and with other stakeholders of the project, leading to more sustainable relationships on the local level with all people directly or indirectly involved with the project.

4.3   SUSTAINABILITY AT THE ORGANIZATIONAL LEVEL

Sustainability and projects have implications for the individual project manager and also for the way we carry out projects. These implications are described in the previous sections. At both levels, there will be implications in terms of the governance of individual projects, the way organizational change skills will develop and the management of the project portfolio.

It is apparent that project managers are well positioned to influence the sustainability aspects of projects, both in terms of the content and the process of the project. However, here are some conditions for this influence to be effective:

- In order for project managers to take responsibility for sustainability, a mind shift of both project managers and project sponsors may be required, in which project managers position themselves as professionals in (sustainable) change.
- Most opportunities to influence the sustainability aspects of the content of the project are during the definition and initiation stage. As professionals in (sustainable) change, project managers should already be involved in projects during these stages, and not just during the execution stage. Indeed, organizations should also involve appropriately gifted and knowlegable project managers in the formulation of portfolio and programmes. Repositioning the project manager will also require a mind shift from project sponsors and portfolio managers.
- The role as professionals in (sustainable) change has implications for the competences of the project manager. Additional competences that are required include consulting skills, awareness of social and environmental aspects and the ability to handle complexity and uncertainty.

4.3.1   Implications at the level of the organization

The conclusions summarized above impact the way organizations organize and govern projects. When considering this impact, we reach the following conclusions.

4.3.1.1 Implication 1: The design and management of the organizational project portfolio should include consideration of the sustainability maturity perspective
A portfolio of projects should be designed from a strategic perspective. This strategy is normally organized around market issues: shareholder value, new markets, new products or services, regulations and internal efficiency and effectiveness of the business processes necessary for an optimal (financial) business result and shareholders' value. Sustainability is emerging as a new strategic direction for companies and will and must have impact on products and business processes. Of course the strategic question is how to translate sustainability into practice in a way that it ensures it remains compliant with the organization's mission and ambition. Clear possibilities come to mind when using the sustainability maturity-levels model as a strategy change model. When the company chooses the resource level as strategic choice, this leads to projects where you seek to eliminate pollution or avoid scarce resources as the basis for products, replacing them with better resources. A choice for strategy at the product level will introduce projects designing new products and deleting existing less sustainable ones. Choices will be derived from market needs and ambition and vision. Combinations of the outlined strategic choices are, of course, also possible. In general, incorporating sustainability creates new criteria (alongside traditional financial measures and shareholder value) for various organizational processes such as aligning strategy and project goals, the discovery of new or different stakeholders, risks, the management of resources, the management of business processes, management control, financial management

and benefits management. The processes (the steps to follow) associated with these aspects will remain similar but the content will differ. Sustainability will lead to different and new stakeholders, risks, environmental and social progress indicators.

4.3.1.2 Implication 2: As a specialist in change, the project manager should be involved during the initiation phase of projects

Traditionally the project sponsor sets the outlines for the project (goals, desired results, funding) where the project manager leads the team realizing the change. It is argued that project managers should take responsibility for sustainability and they are seen as specialists in change. Most sustainability aspects can be best influenced during the definition and initiating phase of the project. Combining the need for taking responsibility, being a specialist in change and the phase in which one can take influence, it's logical to involve (and commit) project managers during the initiation phase of the projects. This changes the process of management control (involving new roles) and the leadership (involving and committing project managers) in terms of the way projects are initiated, directed and terminated.

4.3.1.3 Implication 3: Structural and continuous development of project management competences and the competences of project managers is necessary

The organization needs skilled project managers 'taking' responsibility for sustainability. The process of resource and Human Resources (HR) management covers the management of all types of resources required for project delivery and management. Taking responsibility for sustainability has implications on the expected knowledge and competences for the project manager. These competences should be developed as part of the organizational and personal development plans. They should also be included in job descriptions and should be included in performance appraisal and compensation systems. Delivering skilled project managers also reflects on the hiring processes and requirements of external suppliers of manpower.

4.3.2 Implications for organizational portfolio, programme and project management

The integration of sustainability into the project and on a personal level led us to conclusions in terms of how the organization and governance of the (portfolio of) projects are derived. How should these implications apply to the standards and best practices of portfolio and programme management? As a reference guide for structured best practice, this section analyzes the impact of integrating sustainability on the Portfolio, Programme and Project Management Maturity model (P3M3®). P3M3® describes the portfolio, programme and project-related activities within process areas that contribute to achieving a successful project outcome (See Annex G) (Office of the Government Commerce 2010). The levels described within the P3M3® indicate how key process areas can be structured hierarchically to define a progression of capability which an organization can use to set goals and plan their improvement journey. The levels facilitate organizational transitions from an immature state to a mature and capable organization, with an objective basis for judging quality and solving programme and project issues. P3M3® recognizes not only the project management activities being carried out at the individual project level, but also those activities within an organization that build and maintain a programme and project infrastructure of effective project approaches and management practices.

Section 4.1 already addressed the impact of sustainability at the level of the individual project. And since a programme is considered to be a group of related projects and activities that share a common goal, the analysis in this section is focused on the portfolio of projects. Table 4.5 shows an overview of the implications derived earlier, the effect of these implications on the processes of the P3M3® model and how these implications should be addressed in 'sustainable portfolio management'.

The analysis above shows that integrating sustainability into projects and project management logically also influences organizational portfolio management. Given the relations between project, programme and portfolio management, this impact is not unexpected. But does the impact of integrating sustainability into projects and project management extend even further? Are aspects or elements of the organization effected that are not covered with the concepts of project, programme or portfolio management? For this analysis, the P3M3® model is not sufficient. A more organization-wide model is needed to analyze the full impact on the organization.

4.3.3   Implications for the organization as a whole

A widely used analysis framework that covers the organization as a whole is the European Foundation for Quality Management (EFQM) Excellence Model. The EFQM Excellence Model is a non-prescriptive framework for organizational excellence and maturity. The model is based on nine criteria: leadership; people management; policy strategy; resources; processes; People satisfaction; Customer satisfaction; impact on Society and business results. Figure 4.9 shows an overview of the model. The nine variables are described in Annex H.

The EFQM Excellence Model is used by over 30,000 organizations around the globe in their assessment of organizational excellence and maturity.

Next to the nine criteria, the EFQM Excellence Model identifies five levels of maturity or development of an organization. Each level of the maturity of the organization provides opportunities for improvements, but also implies an

Table 4.5   Implications of sustainability on portfolio management processes

Implications	Impact on P3M3® process	Sustainable portfolio management
Implication 1: The design and management of the organizational project portfolio should include consideration of the sustainability maturity perspective	The process of *management control* describes the activities of starting and stopping projects. It gives no recommendations about the actual evaluation criteria to start/stop projects.	Incorporating sustainability will change the evaluation of progress and alignment with strategic orientation. The evaluation criteria will be organization-specific derived from the application of the six sustainability principles.
	The benefits and financial management processes define the management of both aspects but give no content-driven recommendations including sustainability aspects.	The process of financial management and benefits management itself will hardly alter. Nevertheless, the traditional assessment criteria for directing these projects like finance or other business benefits are no longer applicable enough within this new context. Simply evaluating against financial business case is not compliant with the sustainability principles. The new evaluation criteria, an extended business case, can be derived from the principles of sustainability. This adjusted evaluation framework will be constructed from the principles of 'harmonizing economical, social and environmental', 'short-term and long-term orientation', 'consuming income and not capital', and 'local and global orientation'. Organizations will probably create norms or norm ranges for each of these principles to make the evaluation transparent and repeatable. The principles underlying this framework cover the 'hard' evaluation criteria. The principle of 'transparency and accountability' will address the issue of sharing information within and outside the organization about the project, project results and the effects of the project. The principle of 'personal ethics' will probably not have any impact of the evaluation of individual projects; however a project manager could return or advise to alter or stop an initiative when it does not comply with sustainability strategy. These criteria will be the process-oriented evaluation criteria.
	Stakeholder management	The key impact to the management of stakeholders is the evolvement of new and different stakes, new stakeholders and treatment of long-distance stakeholders. Several aspects of sustainability formulated by the six principles will lead to new stakeholders or to stakeholders obtaining new or different opinions. As a first step in the management of stakeholders is to

Table 4.5   Implications of sustainability on portfolio management processes *continued*

Implications	Impact on P3M3® process	Sustainable portfolio management
		redefine the stakeholders, their interest and they way they influence your company. The principle 'local and global orientation' leads to stakeholders from all over the world but also implies stakeholders very regional/local orientation with huge impact from current media possibilities. The 'both short-term and long-term orientation' principle leads to stakeholders with a very strong long-term orientation or an extreme short-term interest (for example, environmental pressure groups, trade unions and political parties). From the principle 'balancing or harmonizing social, environmental and economical interests' stakeholders can be identified who will take an opposite position to the products or create awareness with consumers about their choices.
	Risk management views the way in which the organization manages threats to, and opportunities presented by, the initiative. There is no explicit or implicit indication of risks associated with sustainability.	The process of managing risk itself will not change. The sustainability strategy will, however, lead to new risks for which (probably more complex) countermeasures should be taken.
	The process of organizational governance views how the delivery of initiatives is aligned to the strategic direction of the organization. It gives no direction of specific sustainability goals or aspects.	In sustainable portfolio management the strategic alignment would take sustainability considerations into account in defining the portfolio for the organization.
	The resource management process covers management of all types of resources required for delivery. These include human resources, buildings, equipment, supplies, information, tools and supporting teams. A key element of resource management is the process for acquiring resources and how supply chains are utilized to maximize effective use of resources.	The process of managing resources itself will not be changed. But there will be a change in the evaluation criteria for acquiring resources and the effectiveness the supply chain. These evaluation criteria come from the sustainability principles. The principle 'balancing or harmonizing social, environmental and economical interests' questions whether the used resources are indeed in balance, the principle 'local and global orientation' will oppose questions from where to obtain resources and the implications from that choice, the principle of 'consuming income and not capital'

Table 4.5   Implications of sustainability on portfolio management processes *concluded*

Implications	Impact on P3M3® process	Sustainable portfolio management
		questions the use of irreplaceable resources like oil and whether replacements or substitutes can be made. These last principles are extra imposed with the principle of 'both short-term and long-term orientation'. Is short-term profit with irreplaceable resources acceptable versus less profit but more durable resources?
Implication 2: As specialists in change, the project manager should be involved during the initiating phase of projects	Management control covers the internal controls of the initiative by a controlling body. It gives no clue who (what roles) should be part of that controlling body.	Based on their role and expertise as specialist in sustainable change, project managers are involved in the controlling body of the portfolio.
Implication 3: Structural and continuous development the competences of projects and the project managers is necessary	Resource management.	In sustainable portfolio management it is expected that project managers are specially trained in sustainability aspects.

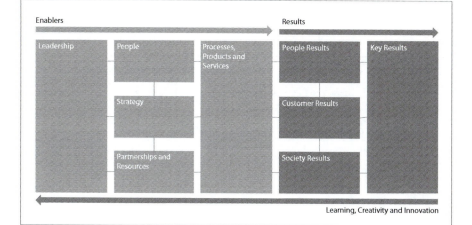

Figure 4.9   Overview of the EFQM Excellence Model 2010
Source: EFQM Excellence Model © EFQM, Brussels, Belgium

indication of the level of change that the organization can handle. Table 4.6 provides a brief description of the EFQM excellence maturity levels.

As stated above, the maturity of the organization also impacts its ability to change. When the EFQM excellence maturity levels are combined with the Sustainable Project Management Maturity model presented in section 4.1.2 you can get a sense of the organizational ability to integrate sustainability in projects and project management. Table 4.7 demonstrates this insight.

Table 4.6   The EFQM excellence maturity levels

Maturity level	Description
Activity oriented	This level is about quality in one's own work situation where everybody tries to do the best job possible.
Process oriented	The distinct steps in the working process are under control. Accompanying tasks and responsibilities are explicitly defined.
System oriented	At all levels people work at continuous improvement. The organizations try to prevent problems instead of resolving them.
Chain oriented	With partners it is tried to gain maximum added value.
Transformational oriented	The strategy is focused to become at the top performing companies

Table 4.7   The EFQM excellence maturity levels confronted with the levels of consideration of sustainability in projects and project management

		Level of consideration of sustainability in projects and project management			
		Resource	Business projects	Business model	Product/service
EFQM Maturity levels	Activity oriented	Yes	No	No	No
	Process oriented	Yes	Yes	No	No
	System oriented	Yes	Yes	Possibly	Possibly
	Chain oriented	Yes	Yes	Yes	Yes
	Transformational oriented	Yes	Yes	Yes	Yes

From this analysis, a number of observations can be drawn:

1. It should be easy to apply sustainability at the resource and business process level. This is logical because these levels are more about optimizing than about improving. On these levels, there is no real need

for a breakthrough in thinking. When there is insight in the integral process and people understand that success (delivery process) is only possible when they cooperate, changes are easier to implement.

2. Changes in terms of the business model of products and services are only possible within a mature organization.

3. Within a system-oriented organization, the culture and people will determine whether it is possible to improve at the business model level or product/service level. A collaborative culture will have a positive effect on these possibilities.

From this analysis it shows that organizational maturity is also a measure of the ability of the organization to initiate change. This ability or maturity should be considered when defining projects and the portfolio of projects.

REFERENCES

Ashkanasy, N.M. and Ashton-James, C.E. (2007), Positive Emotions in Organizations: a Multi-level Framework, in Nelson, D.L. and Cooper, C.L. (Eds), *Positive Organizational Behavior*, Sage, London, pp. 214–224.

Cameron, K. (2008), *Positive Leadership: Strategies for Extraordinary Performance*, Berrett-Koehler Publishers, San Francisco, CA, USA.

Carnegie Mellon Software Engineering Institute (2002), Carnegie Mellon Software Engineering Institute: Capability maturity models, Retrieved 29 Decemberm 2010. Available at: http://www.sei.cmu.edu/cmmi/.

Cerny, A. (2009), Management of Emotions in Projects, in Kähköhnen, K., Kazi, A. and Rekola, M. (Eds), *Human Side of Projects in Modern Business*, IPMA Scientific Research Paper Series, Helsinki, Finland.

Clifton, D.O. and Harter, J.K. (2003), Investing in Strengths, in Cameron, K.J., Dutton, J.E. and Quin, R.E. (Eds), *Positive Organizational Scholarship: Foundations of a New Discipline*, Berrett-Koehler Publishers, San Francisco, CA, USA.

Dinsmore, P.C. (1998), How grown-up is your organization? in *PM Network*, 12 (6): 24–26.

Eid, M. (2009), *Sustainable Development & Project Management*, Lambert Academic Publishing, Cologne.

Emmons, R.M. and Cullough, M.E. (2003), Counting blessings versus burdens: experiment studies of gratitude and subjective well-being in daily life, in *Journal of Personality and Social Psychology*, Vol 84(2), Feb 2003, 377–389.

Folkman S. (1997), Positive psychological states and coping with severe stress, in *Social Science Medicine* Volume 45, Number 8, October 1997 (45), 1207–1221.

International Project Management Association (2006), *IPMA Competence Baseline version 3.0*, International Project Management Association, Nijkerk, the Netherlands.

Knoepfel, H. (Ed.) (2010), *Survival and Sustainability as Challenges for Projects*, International Project Management Association, Zurich.

Losada, M. and Heaphy, E.D. (2004), Positivity and connectivity in the performance of business teams, in *American Behavioral Scientist*, Vol. 47 No. 6, February 2004: 740–765.

Morris, D. and Garrett, J. (2010), Strengths: Your Leading Edge, in Liney, P.A., Harrington, S. and Garcea, N. (Eds), *Oxford Handbook of Positive Psychology and Work*, New York: Oxford University Press.

Office of the Government Commerce (2009), *Managing Successful Projects with PRINCE2*, Norwich, TSO.

Office of the Government Commerce (2010), *Portfolio, Programme and Project Management Maturity Model (P3M3®): Introduction and Guide to P3M3®*, Norwich, TSO.

Prat, M.G. and Ashforth, B.E. (2003), Fostering Meaningfulness in Working and at Work, in Cameron, K.J., Dutton, J.E. and Quin, R.E. (Eds), *Positive Organizational Scholarship: Foundations of a New Discipline*, San Francisco, CA, USA, Berrett-Koehler Publishers, pp. 111–121.

Project Management Institute (2007), *Project Manager Competency Development (PMCD) Framework*, Second edition, Project Management Institute, Newtown Square, PA, USA.

Project Management Institute (2008), *A Guide to Project Management Body of Knowledge (PMBOK® Guide)*, Fourth edition, Project Management Institute, Newtown Square, PA, USA.

Project Management Institute (2010), *Code of Ethics and Professional Conduct*, Project Management Institute, Newtown Square, PA, USA.

Seligman, M.E.P., Steen, T.A., Park, N. and Peterson, C. (2005), Positive psychology progress; empirical validation of interventions, in *American Psychologist*, Vol 60(5), Jul–Aug, pp. 410–421.

Silvius, A.J.G. (2010), Workshop Report Group 2, in Knoepfel, H. (Ed.), *Survival and Sustainability as Challenges for Projects*, International Project Management Association, Zurich, pp. 155–160

Silvius, A.J.G. and Schipper, R. (2010), A Maturity Model for Integrating Sustainability in Projects and Project Management, International Project Management Association, 24th World Congress, Istanbul.

CHAPTER 5

REFLECTION AND CONCLUSION
RON SCHIPPER, GILBERT SILVIUS AND JASPER VAN DEN BRINK

The previous chapters demonstrated the logic of incorporating the principles of sustainability in projects and project management and analyzed how this would impact the most common standards of project management. This analysis was thorough, specific and precise. But the question remains: what's new? Project managers deal with interests of diverse stakeholders every day, so adding a few new considerations and perspectives may be challenging, but no more than that. Does integrating the principles of sustainability change the profession of project management? Or is it in essence 'business as usual'?

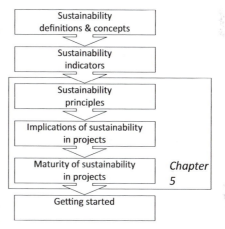

This chapter reflects upon these questions. We will discuss what's really new about integrating sustainability in projects and project management. We will show that taking responsibility for sustainability is about professionalism and ethics as a project manager. We will elaborate on the principle 'Personal values and ethics' by discussing some of the dilemmas that ethical behaviour include. The final section of this chapter provides the conclusion of this book and shows that project managers not only *should* take action, but also *can* take action.

5.1   SO, WHAT'S NEW?

In the introduction of this book, we asked ourselves the question 'What's new?' What does the integration of sustainability in projects and project management really change about project management?

Chapter 4 analyzed the impact of the principles of sustainability on projects and project management. From this analysis it became clear that these principles, particularly those such as 'Harmonizing social, environmental and economical interests', 'Both short-term and long-term orientation' and 'Local and global orientation' provide new or additional perspectives on the content and the process of the project. So we may conclude that integrating sustainability requires a *scope shift* in the management of projects; from managing time, budget and quality, to managing social, environmental and economic impact.

However integrating sustainability in project management is more than adding a new perspective or aspect. It is more than changing some processes and some formats within the current project management standards. Adding new perspectives to the way projects are considered also adds complexity. Chapter 3 made the case that organizational change is a more complex level of change than the change resulting from traditional (building and construction) projects. Project management therefore needs a more holistic and less mechanical approach. The traditional project management paradigm of controlling time, budget and quality suggests a level of predictability and control that is just not realistic in complex changes. The integration of sustainability requires a *paradigm shift* – from an approach to project management that can be characterized by predictability and controllability of both process and deliverables, and is focused on eliminating risks, to an approach that is characterized by flexibility, complexity and opportunity. Figure 5.1 illustrates this paradigm shift.

Traditional		**Sustainable**
Project Management		**Project Management**
Time, Budget, Quality	——————→	Social, Environmental, Economic
Inside-out	············→	Outside-in
Shareholder	——————→	Stakeholder
Process	············→	Content
Threats	——————→	Opportunities
Path	············→	Steps
Control	——————→	Guide
Knowing	············→	Learning
Output	——————→	Outcome
Closed	············→	Open

Figure 5.1   The paradigm shift of sustainable project management

The basis for the shifts described above is the way the project management profession sees itself. Traditionally, project managers tend to serve their project sponsors and 'do what they are told' to do. They position themselves as subordinate to the project sponsor and manage their project team around scope, stakeholders, deliverables, budget, risks and resources as specified by the stakeholder's requirements. The core message of this book is a call to action to the project management profession to take responsibility for a more sustainable development of organizations and business. Taking up this responsibility changes the role of project managers and therefore changes the profession. Integrating sustainability requires that project managers develop themselves as specialists in sustainable development and act as partner of and peer to stakeholders. In this *mind shift* the change a project realizes is no longer a given nor exclusively the responsibility of the project sponsor, but also the responsibility of the project manager with ethics and transparency as a basic touchstone. Project management is no longer about 'managing' stakeholders, but about engaging with stakeholders in realizing a sustainable development of organization and society. Figure 5.2 illustrates this mind shift. To the relation between projects and the permanent organization (as illustrated in Figure 3.2), we add the link between project management and the goals of the organization.

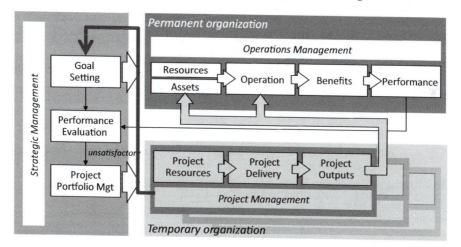

Figure 5.2   The extended relation between projects and the permanent organization

The three shifts identified above of scope, paradigm and mind can be summarized and connected as illustrated in Figure 5.3. The scope shift resides in the paradigm shift of sustainable project management. This paradigm shift is grounded in the mind shift.

The question arises how to apply these in practice?

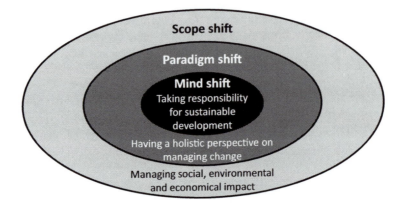

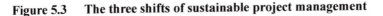

Figure 5.3   The three shifts of sustainable project management

5.2   THE SCOPE SHIFT

As stated, the scope shift is about adding additional viewpoints on projects and project management; from managing time, budget and quality, to managing social, environmental and economic impact. In a certain way it is about applying new knowledge. This is formulated in the six principles of sustainability and the project management checklist making the principles more practical. Chapter 4 gives guidance on how to apply this knowledge to the individual, project and organizational level. Elaborating on this, Chapter 6 is dedicated to the different project roles where concrete first steps are defined for getting started.

5.3   THE PARADIGM SHIFT: EMBRACING UNCERTAINTY

The paradigm shift consists of accepting and handling complexity and uncertainty instead of ignoring or simplifying it by both project manager and project sponsor. And although the inadequacy of traditional project management as a technique to manage organizational change was already addressed in section 3.1, the scope shift illustrated in the previous section, and the increased complexity that comes along with this, emphasizes the need for a new approach. The desire to consider organizations as controllable systems that can be changed top-down and in a predictable way is deeply ingrained – from investors and financial markets, to executive boards and project sponsors. It is therefore not easy to accept uncertainty as an inextricably linked part of complex changes and projects. Nevertheless, in reality, complexity and uncertainty are there and need to be handled.

A popular way of coping with uncertainty is to transfer it, together with the risk related with it, to the project organization and more specifically, to the project manager. In project plans, the deliverables, quality requirements and budgets are fixed, and uncertainty is handled with the change management and risk management processes.

The fact that only a small fraction of projects actually delivers according to specifications, on time and on budget (as reported in many studies, the most famous one being the Chaos Reports of the Standish Group 1995 and 2009), shows the inadequacy of this approach. Project managers are not in a position, and do not put themselves in a position, to handle real world uncertainty because this usually involves changing the specifications of the deliverables of the project, the time schedule and/or the budget. And the project sponsor, who is in the position to make these kind of decisions, usually keeps their distance from the project, because they feel they hired the project manager to 'do the job'. It is therefore up to the project manager who has then to defend a project planning and project budget configured at the initiation of the project (months or years ago).

How can real world uncertainty be handled and not ignored? The key to this question is in choosing an approach that *embraces uncertainty and complexity*. And this approach calls for an *iterative way of doing projects*, with *active involvement of stakeholders* and *frequent dialogues* with all participants. In this approach, project managers are managers of change, that focus on taking small (short in time) *iterative project steps* (iterations), each time *delivering business value* as soon of possible. They manage on *team productivity*, instead of the golden triangle constraints of time, budget and quality.

Most of the principles are found in the methods of complex project management (see Chapter 3) and Agile-development of software (See Annex I). The Agile-development manifest refers to 12 principles. The most relevant principles are:

1. *Welcome changing requirements, even late in development and regular adaptation to changing circumstances*
   Where there is severe complexity and uncertainty changes are inevitable. The Agile approach is able to cope with these late changes without compromising capita or business revenue. In the project management checklist, the more concrete application of the sustainability checklist, flexibility and optionality in the project is proposed. The Agile approach realizes a structure for doing that in the project delivery process.
2. *Sustainable development, able to maintain a constant pace*
   The Agile manifesto states that the Agile project approach is designed in a way that sponsor, engineer and end-users can follow this process endlessly. The Agile approach applies the sustainability principle 'of consuming income, not capital' in a practical way.
3. *Close, daily co-operation between business people and developers*
   In a complex world there is no longer one person able to define every

solution; so the best way is to join together and give from your role ideas to solutions or next steps.

4. *Projects are built around motivated individuals, that are trusted and self-organizing teams*

   These principles refer to a positive treatment of people, which is reflected in the project management checklist. It's a sustainable conviction that motivated people don't need to be controlled or a project manager has to think for them, but the project manager should help and create the circumstances needed for them to be able to do the right (sustainable) thing.

5. *Project (sub)results are delivered frequently (weekly rather than monthly) and real project (sub)results is the principal measure of progress*

   This principle is within the Agile manifesto seen from a 'working software' perspective, but can also be viewed from a more general project result viewpoint. It refers to small short time iterative steps in the project delivery approach. Complexity and uncertainty can be better overviewed in small iterations whereas each iteration delivers the organization will a useable (sub)result. With these (sub)results, the productive phase of the project deliverables can start straight away. And when later in time the produced (sub)result turns out to be not suitable anymore, it can be altered or replaced. The costs of doing that are less compared to building the whole result at once and also delivering it later.

It should be concluded that these project management approaches are more suitable for handling real world complexity and uncertainty in projects.

5.4   THE MIND SHIFT: SUSTAINABILITY, PROFESSIONALISM AND ETHICS

In the introduction to this book we quoted Mary McKinlay's opinion that 'the further development of the profession requires project managers to take responsibility for sustainability' (McKinlay 2008). With this statement, McKinlay connects sustainability to the development of the project management profession. An interesting view, but given the conclusion of Eid's study, project management fails to address the sustainability agenda, one that may not necessarily be shared by many professionals in the field (Eid 2009). So let's take a closer look at what makes a profession a profession and how this may relate to sustainability.

Amongst both practitioners and academics, there seems to be a general consensus that project management is moving towards a 'true' profession (Jones and Young 2007). A profession in the tradition of a lawyer or a doctor. A profession as opposed to an 'occupation'. Motivations for this development, however, may differ. Practitioners seem to join professional associations and meet with like-minded people to discuss the development of their occupation in a way which will lift the standard of service provided to their customers. Academics seem to be more concerned with correct use of terminology and what that may mean for further theoretical development. Jones and Young believe that with these 'opposing views, the academics attempt to theorize and rigorously test the semantics of the argument, whilst practitioners are more interested in how status applies "at the coal-face"'.

However, the development of project management as a profession is not undisputed. For example Zwerman et al. (2004) conclude, 'Project Management is not now, nor is it likely to be considered a profession in the foreseeable future.' The discussion on what the position of project management is on the scale of 'occupation' to 'profession' brings up the question: What determines a profession?

Several publications have discussed the question what determines a profession. In these publications a profession is typically defined as, 'A calling, requiring specialized knowledge and often long and intensive preparation including instruction in skills and methods as well as in the scientific, historical, or scholarly principles underlying such skills and methods, maintaining by force of organization or concerted opinion high standards of achievement and conduct, and committing its members to continued study and to a kind of work which has for its prime purpose the rendering of a public service.' A profession implies systematic knowledge and proficiency. This relationship is further developed by Boone (2001) who states that, 'Professions are based on scientific and philosophical facts acquired through scholarly endeavour.'

However a well developed knowledge base is not the only characteristic that defines a profession. From a diverse set of publications, the following set of criteria for the recognition of a profession can be developed.

- *A profession must have a well developed and theoretically sound body of knowledge*
  An impressive number of books have been written on the subject of project management. Accumulated, these publications document the body of knowledge that exists in the project management community. However, publication and methods on project management in general lack a strong theoretical foundation (Koskela and Howell 2002). In the underpinning of the project management body of knowledge, a discrepancy between academic practice and business practice shows. Project management is often considered underdeveloped as an academic field of study.
- *A profession must have a high level and specialized education and training structure*
  The discrepancy between academic and business practice also shows in this criterion. The private sector has a well-developed training and education

market, but at universities substantial project management education hardly exists. For further professionalization it is of crucial importance that project management is recognized as a professional, but also as an academic discipline.

- *A profession must allow professionals to develop their career in the profession*
  Not just in education and training should project management be recognized as a profession, but also in career development in organizations. Today, it is mostly in professional services companies and project industries that project management can be chosen as a career step or career path. In other industries, like the public sector, manufacturing, financial services, healthcare, and so on, project management is still considered an 'accidental job' that other professionals do as an assignment (Pinto and Kharbanda 1995).

- *A profession must have a high-level image in society and ditto social status*
  An important part of being a profession is to be regarded as one. In this aspect the work of Paul Giammalvo is interesting (Giammalvo 2007). He studied the perception of project managers compared to other professions. In his study, project management still scores far from traditional professions like doctors or lawyers. For a relatively young profession this should not be surprising, but it also means that there is still a lot of work to do on the perception of project management. We believe that connecting project management and sustainable development will benefit the perceived contribution that the project management profession makes to society.

- *A professional must have a substantial level of professional autonomy in his work*
  This criterion says that a profession cannot exist if all the actions that the person has to do in this 'profession' are more or less prescribed. Being a profession implies that the professional has to be able to make decisions in his work, based on his 'professional' judgement and consideration of the circumstances. This criterion emphasizes the call to action this book includes. Project managers should demonstrate an autonomous professional responsibility for the projects and the change they realize.

- *A work of a professional must be relevant to society and add to its wellbeing or prosperity*
  This criterion focuses on the contribution that a profession makes to society. Basically it asks: does the profession matter? We believe that project management scores high on relevance, at least measured to economical standards. As stated earlier, organizations need to adapt to changing circumstances and environments in order to stay competitive and projects are the preferred way of organizing this change. This development results in billions of dollars or euros being spent on projects, with estimations up to roughly 25 per cent of Gross Domestic Product (GDP) (Turner et al. 2010). The economic relevance of projects and project management is therefore

strongly proven. With regards to the other pillars of sustainability, social and environmental, the relevance of project management is most of all represented by the changes that projects generate.

- *A profession must have a code of conduct and regulatory body with some level of sanction power*
  Another characteristic of a profession is that it is legally protected. This means that only persons matching certain criteria can carry out the profession and that it organizes its own subsystem of professional jurisdiction, within the general laws that apply to everyone. Strong examples can be found in the medical profession and the legal profession. As illustrated earlier, the professional bodies for project management also have codes of conduct. However, it should be concluded that market recognition of these codes is still limited. A significant development in this area could be the steps the British Association for Project Management (APM) is taking to become a chartered body. APM does so in order to raise standards amongst the project management community.

In our search for the connection between sustainability, professionalism and ethics, the last three criteria of the list above need further examination. From these criteria, it should be concluded that being a true profession suggests that, in exercising their profession, professionals have an autonomous and proactive approach to their role and perform this role with awareness of the role of their profession in society. Given the relationship between sustainable development and projects demonstrated in section 1.2, this calls for the project management profession to contribute to sustainable development in society. *The further development of project management as a profession therefore requires project managers to take responsibility for sustainability.*

This professional autonomy of the project manager also implies responsible behaviour towards society. And since ethical behaviour can be seen as a part of responsible behaviour, it should therefore be concluded that *the further development of the project management profession also requires project managers to consider the ethical aspects of their work.* In fact, the *PMI® Code of Ethics and Professional Conduct* also connects 'ethics' to 'professional conduct' or professionalism. In section 3.1 we showed that the *PMI® Code of Ethics and Professional Conduct* connects sustainability and ethics. Should we thus conclude that these three terms or concepts are interrelated? We believe we should. *The three concepts are intertwined and cannot be considered in isolation.*

5.5   ETHICS OF THE SUSTAINABLE PROJECT MANAGER

The previous section connected sustainability to the profession of the project manager and to ethics. *But what is ethics? What is ethical behaviour? And how can you apply it?* The following section aims to give project managers some insight and guidance in ethical behaviour and moral judgements. It provides a brief introduction into the topic, followed by two alternative approaches to ethical dilemmas. It also provides a method to deal systematically with these dilemmas.

5.5.1   Ethical dilemmas

Working as a sustainable project manager is not easy. The theory is clear but in practice balancing social, environmental and economic interests is a Herculean task. The simple scenarios are easy to deal with. For example, child labour is a definite no-go. But what do you do if all the competitors use child labour and the project will be too costly when you're the only one working sustainably? What if this leads to bankruptcy for the organization? Which means loss of jobs and loss of well-being? Perhaps child labour is still a no-brainer, but how about bribery? In many cultures, bribery is generally accepted and a fact of life. Now things are getting more complicated.

The same dilemmas hold true for many other sustainable decisions. Ecological meat production implies a better quality of life for the animal. At the same time it may require more energy use and pollution, not least because the animal is using his energy to walk around instead of gaining weight.

And how about a project to outsource the call centre activities to India? It will create jobs in India, which is a good thing, but at the same time people at home may lose their jobs. In other words a project manager is often faced with difficult and moral decisions in which he or she has to choose between various competing options.

In practice most people are not faced with major ethical dilemmas that often. In many cases it is simply enough to look for the option that is most profitable and legal. And yet a project manager cannot hide behind a single focus in terms of profitability. He or she needs to consider all the principles of sustainability and these can easily lead to dilemmas. Ethical dilemmas can be described as problems for which there are no solutions that do not, in turn, create problems for others as a consequence.

5.5.2   Codes of ethics or conduct

Project managers can get some guidance in ethical dilemmas from ethical codes, with the *PMI® Code of Ethics and Professional Conduct* as the most obvious one to use. The *PMI® Code of Ethics and Professional Conduct* aims to give a set of values and guidelines that should lead project managers in making ethical choices. The code identifies honesty, responsibility, respect and fairness as leading values.

A problem with codes of conduct is that the general principles may be logical and clear but the application of these principles in day-to-day decisions remains open for interpretation. It is therefore important to develop real-life examples that shed more light on how the ethical and professional guidelines should be interpreted, in the context of real-life ethical dilemmas. However, based on a poll of International Project Management Association (IPMA), Project Management (PMI) and (Australian Institute of Project Management (AIPM), Giammalvo (2007) concludes that 'these professional organizations are generally not aggressive in enforcing codes of ethics or codes of conduct, especially for violations which impact the consumer of their services'.

A big advantage of a code of conduct is the transparency of the process. With a code of conduct is it is clear what kinds of criteria are being used in order to judge decisions. This section provides supplementary information on the process of moral decision making, without claiming to make this kind of decisions ever easy. Using a code of conduct makes these kinds of decisions easier.

Making ethical decisions is about taking into account the interest of other people. Making sustainable choices asks for even more, since sustainability implies stakeholders who may be nearby or across the other side of the globe. Stakeholders include people who are not yet born, but will live on this planet in the next hundred years. This leads us to a fundamental question: How can we take their interest into account? And especially the interest of those whose opinion we cannot ask? Using a utilitarian and a deontological perspective can help us to solve this problem.

5.5.3 Utilitarian ethics: look at the consequences

A first perspective we can use in answering this question is that of utilitarian ethics. Utilitarian ethics is based on the assumption that choices should be judged on the consequences (consequentialism). When you are in moral doubt you should evaluate the consequences of all feasible actions. An approach that is often used in our day-to-day decision-making processes. We are often looking at the consequences of our decisions, using rules of the thumb, previous experiences and sometimes the knowledge provided by science.

In making this evaluation, consequentialists look at the consequences for different actors. A project manager can look at the consequences for themselves, but also to the consequences for people in their immediate vicinity, or then other people who will face the consequences of their action, for example children in the developing world or future generations.

This is one of the difficulties of consequentialism. It is very difficult to predict the future. This makes a theoretically easy method of making decisions a far more complex decision process in reality. For example, who can give a good prediction of the developments in nuclear waste handling in the next 50 years? Very few people (if any) can, which leaves a project manager who uses a utilitarian decision-making process empty handed on one of the most important decision criteria when it comes to using nuclear energy.

Another problem is the standard metric that is used when a utilitarian decision is made. What's the goal? Are we trying to maximize wealth for as many people as possible? Or do we evaluate the consequences in terms of the amount of pleasure or pain to people? Or is the environmental health of our planet the most important norm? When using different norms to assess the consequences of a decision you will arrive at different outcomes. Utilitarian ethics tells that a consequentialist choice is right if it maximizes the usefulness for as many people as is possible. Which leads to the question: when do the rights of the many outweigh those of the few? According to John Stuart Mill (1862) there is a limit to utilitarianism, whatever the utilitarian choices you should always respect the freedom of the individual. In this line of reasoning killing a few to make the rest of the population happy is morally wrong.

5.5.4   Deontological ethics: use moral 'laws' that everyone can agree on

A second perspective is the Deontological approach, introduced by Immanuel Kant in his influential *Metaphysics of Morals*. (1797) Kant's primary goal in ethics is freedom: we should not be slave to our tradition or our senses. Kant's primary goal in ethics is freedom: we should not be slave to our tradition or our senses. Freedom can be reached if our moral choices are founded on universal moral laws, called maxims.

Kant states that a maxim should indicate how to choose in situations we have to deal with. To make an effective maxim is complicated; an important criterion for a maxim is the extent to which it will hold up as a universal natural law. If a moral maxim can act as a natural law, everyone can act according to it, so it can be used as a moral law on which to act.

5.5.5   How to use the deontological and utilitarian perspectives in practice

When using the deontological try the following simple experiment (based on the veil of ignorance approach of John Rawls (1971)). Imagine that you do not know where you live (in the USA, in Bangladesh or in China) and that you don't know the colour of your skin, your sex, your age or your occupation. What would be the best norms to use for this moral decision?

When it comes to sustainability you need to add an extra question. Sustainability is not only about other people on the planet, but is about the future generations as well. So imagine that not only don't you know where you're living, you also don't know when. To make it more practical: can you explain to your children and grandchildren the choices you made and would you make the same choices if you were in their position?

This experiment forces you to take various positions that are all relevant for the moral decision that is being made. It helps you search for fundamental norms on which to build the moral decisions. Which fundamental norms will be valid for all the different parties involved?

When using the utilitarian approach, the main question is to look at the end result for all parties involved. With this approach all relevant actors are taken into account, not to find a fundamental norm or maxim, but in order to survey the different consequences for the parties involved.

5.5.6 Moral reasoning in steps

There are many ways to choose moral responsibility. We will present an approach that can be used to make a step-by-step decision. To this end, we divide the moral decision-making process into three phases; information gathering, ethical analysis and the actual decision making and evaluation.

Step 1. Collect the facts around the moral dilemma you are in

The first action is to describe precisely what the moral problem is. By clearly describing what the problem is you are able to keep focus during the other phases of process. So be aware of the scope of this small project!

The action step is to collect all the relevant facts. For a moral decision it is important to know who the stakeholders are, what their interests are and what the values are they use to morally defend their interests. A list can be made of all these elements. It is important to make a difference between facts and values. Important in this phase is the willingness to search for information that is conflicting with one's own beliefs. Collect information about the conventional norms or laws that have to be observed as well.

The third action is to consider all possible actions. To be able to make a choice you need at least two possible actions. If there is only one alternative, there is no (moral) problem.

Step 2. Start an ethical analysis

In Step 2 the problem is analyzed using the different moral approaches as described above.

First, from a utilitarian perspective: what are the cost and benefits (the consequences) of the various options? Of course you should not limit yourself to financial costs and benefits. The costs for people and planet need to be taken into account as well. Make a distinction between local and global costs and long-term and short-term costs to make sure that you cover all dimensions.

Second, from the deontological perspective: what are the fundamental norms that need to be observed? When you look at the different values of the different stakeholders, which values could be universal from a rational perspective?

Third, check whether a solution can be found by talking to all stakeholders involved. When this is possible this will lead to a decision with a high level of support from the relevant actors.

Step 3. Decide and look back

In the last step the actual decision is made, based on all the information and insights gathered so far. What seems to be the right decision? And are you able to explain this decision to all stakeholders involved in a way you find satisfying?

This is the evaluation phase as well. What does our intuition tells us? Does this really look like the right decision? Can we look ourselves in the eye and be content with the decision? Is it the right decision from the perspective of the various actors, when we perceive them as reasonable persons? And for those project managers that want to manage a real sustainable project an extra question shows up: have I done everything that is possible to realize the maximum sustainability potential of this project? If the conclusion from this step is a 'no', we should very carefully examine what the problem with the decision is. With this knowledge we should go back to the first step and redo the whole process.

When using these steps we have a procedure to deal with moral problems. We don't have a recipe for a quick and easy solution. Making these kinds of decisions isn't an easy or quick process – it takes time, effort and willingness to solve these problems. But for a sustainable project manager knowing how to deal with these problems is an indispensable competence.

5.6   CONCLUSION

In this book we have shown that projects and sustainability are interrelated. Sustainable development needs us to do things differently; needs change and needs projects to deliver this change. We have also shown that project managers are well positioned to influence the sustainability aspects of the projects they are working on, both regarding the content of the project and the process of the project.

Taking responsibility for these sustainability aspects is actually part of the further development of the project management profession.

The core message of this book is to call to project managers to take responsibility for a more sustainable development of organizations and business. In Chapter 4 we demonstrated how, at various levels, initiative can be taken. We explained the necessary competence at the personal level as well as those aspects over which a project manager has influence or can take full responsibility. At the project level a maturity model provided a central framework for incorporating sustainability into projects. At the organizational level we explored options in terms of governance, leadership and other aspects. Nevertheless, taking the first steps is always difficult. The remaining question therefore is: how to take these first steps? *How can project managers, and other roles in and around projects, take responsibility?*

The next and final chapter of this book is devoted to showing how this responsibility can be picked up, by individual roles participating in a project. It shows in simple steps what he or she can do to get started.

REFERENCES

Boone, T. (2001), Constructing a profession, in *Professionalization of Exercise Physiology (PEP)* online, 4 (5).

Giammalvo, P. (2007), *Is project management a profession? If yes, where does it fit in and if no what is it?*, PhD Thesis, ESC Lille, France.

Jones, D., Young, W. 'Turf Wars – Is project management a profession?' Australian Institute of Project Management National Conference 2007 Hobart, Australia 5–8 October 2007

Koskela, L.J. and Howell, G. (2002), The Underlying Theory of Project Management is Obsolete, PMI Research Conference, Project Management Institute, Newtown Square, PA, USA.

Mill, John Stuart (1862), T*he Contest in America. Harper's New Monthly Magazine*, Volume 24, Issue 143, page 683-684. Harper & Bros., New York, April 1862.

Pinto, J.K. and Kharbanda, O.P. (1995), Lessons for an accidental profession, in *Business Horizons*, March–April.

Project Management Institute (2010), *Code of Ethics and Professional Conduct*, Project Management Institute, Newtown Square, PA, USA.

Rawls, J. (1971), *A Theory of Justice*, Harvard University Press, Cambridge.

The Standish Group (1995), *Chaos Report*, The Standish Group, Boston.

The Standish Group (2009), *Chaos Summary 2009*, The Standish Group, Boston.

Turner, R., Huemann, M., Anbari, F. and Bredillet, C. (2010), *Perspectives on Projects*, Routledge, Abingdon.

Zwerman, W., Thomas, J., Haydt, S. and Williams, T. (2004), *Professionalization of Project Management: Exploring the Past to Map the Future*, PMI Publishing, Newtown, PA.

CHAPTER 6

GETTING STARTED
RON SCHIPPER AND GILBERT SILVIUS

The previous chapters gave insight into the need and opportunities for the integration of sustainability in projects and project management. Now, it´s time to get started. This final chapter provides specific and practical suggestions for a step-by-step approach.

Of course, the different roles in and around projects carry different responsibilities. The suggested steps to 'get started' are therefore structured in these different roles. The main roles in and around projects, project manager and project sponsor, are derived from the analysis in

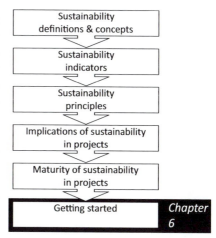

paragraph 4.2, but three new roles are added: project management office (PMO) leader, project user and portfolio manager. These roles are added because they are also very well positioned to influence the sustainability aspects of projects.

The steps for each role are entirely described from the perspective of this person. Of course, in reality, the roles interact on the sustainability aspects and many other aspects and a joint approach may be constructed, to serve the goals and ambition of the specific project or organization at hand.

6.1   GETTING STARTED AS A PROJECT MANAGER

Step 1: Start! Decide to take responsibility for sustainability

As shown in Chapter 4, a project manager always has influence over   the sustainability aspects of their project. For some parts he or she can take full responsibility; for other parts they have at least influence through the project sponsor. Traditionally, project managers wait for the project sponsor to set directions, but key success factor the project manager involves taking action and not waiting for the organization (for example, the project sponsor) to be ready. Of course, some actions require agreement from the project sponsor or general management of the organization. Therefore the initial steps for the project manager can be separated into actions to be taken within the project team and actions to be discussed with the project sponsor.

Step 2: Understand the company

It is important to understand the company's overall mission and ambition with respect to sustainability. This outlook will set the sustainability goals and boundaries for other departments and the projects which run. The project manager can start collecting information from the company by checking their website(s), annual reports and corporate social report (CSR-report) looking for sustainability statements.

Projects are executed within one or more specific divisions or departments. This means that a general view on sustainability will be more focused in each specific department in terms of their customers, individual products, business processes, product lines, culture, behaviour and individuals. The company's vision will have an impact on the specific department. Projects are executed in this setting so the project manager must have a conceptual picture of the direction which the department is taking. For example, the Shell Oil company statement declaring: 'In the next 50 year's governments, energy producers and users should jointly do three things at once: meet the demand for energy, maintain the stocks and limit their impact on the environment and society' [www.Shell.nl]. The last part of this statement specifically will have large impact on the design and execution of the oil related production process. Executing a project with respect to the production process should primarily take this statement into account and the project manager should be fundamentally aware of it.

Step 3: Where should the project be? Set goals for the project

Understanding the organization's ambition (step 2) enables the project manager to formulate the goals with respect to sustainability in their project. Of course the information about the desired and current situations (step 5) and information about the discrepancy between the project and its environment will be used to create a development plan (step 6). The relevant activities in this goal setting phase are:

- identify the stakeholder whose commitment is necessary for this goal setting;
- complete the 'desired' situation of the sustainability maturity model with the help of the project team and stakeholders;

- communicate and discuss this agreed 'desired' situation with the relevant stakeholders.

Step 4: Where is your project now? Analyze the current application of sustainability in a specific project

Once you understand company sustainability at the formal level, the project manager should take a look at the project. Collect all relevant data about the application of sustainability in the project. Then relate this to the context of the department and the company to construct the current situation and give it a numerical value (or assessment).

- Take the project management sustainability checklist from Chapter 3. Which elements of the checklist are currently applied in the project and in which way?
- In what ways are the dilemmas, represented by the six principles of sustainability, tackled in the project? What choices have been made implicit or explicit in the project?
- Complete the project maturity assessment. What is the current level of assessment of sustainability maturity? At what level of vision is the project executed?
- Discuss the findings with the relevant stakeholders.

Step 5: Find and analyze discrepancies

Discrepancies can be found at two levels: the alignment of the project with its environment and the mismatch between the desired and current situation of the project. Both angles should be viewed and analyzed in order to assure the project is contributing effectively. Do not forget that an ambitious project can also change the company's ambition! The analysis leads to the discovery of positive and negative discrepancies. A positive discrepancy means that the project is taking sustainability further than the department or company needs at this stage. A negative discrepancy means that the project scope, approach, results or effect should be changed to address shortcomings.

You need to undertake the following activities:

- consider each fact to assess whether it's a positive or negative discrepancy. Explain how and what the consequent effect is on the project, the department/ company and the customer/society;
- discuss the requisite effort needed to overcome the discrepancy;
- consider the facts and the discrepancies with respect to the project or company as a basis for the subsequent decision process;
- decide which points needs to be changed within the project;

- decide, in discussion with stakeholders, where the ambition statements of the organization or department need to be adjusted;
- agree a final decision with stakeholders.

Step 6: Assess the complexity and uncertainty of the project

As shown in various chapters, the future of projects and project management is the handling of uncertainty and complexity. Therefore, assess the complexity and uncertainty of your project and chose a corresponding approach to handle this in an appropriate way. For example, implement the principles of Agile development in your project to deliver quick usable business results with real value instead of late big business functionality no longer aligned with the strategic orientation.

Step 7: Make a development plan

Logically, the last step is to develop a plan to realize the necessary changes. This plan needs to be a joint effort between stakeholders and the project team.

6.2   GETTING STARTED AS A PROJECT MANAGEMENT OFFICE LEADER

More and more companies organize their responsibility for project management resources in an expert pool. Some organizations also organize the responsibility for project delivery in this same department. In this section the focus is on the project management expertise and not on the project delivery responsibility.

The project management pool in many cases acts as a centre of expertise on project management, then provides the project managers required for various projects in the organization. This development creates a new role in an organization, that of the 'PMO leader'.

This leader can have any combination of the following four responsibilities:

1. develop, maintain and control the standards and methods for projects and project management;
2. supply the organization with skilled project managers;
3. create an environment in which projects can be successful;
4. responsibility for the project delivery of the portfolio of projects.

6.2.1   Embedding sustainability in the project management standards and methods

The standards and methods for project management consist of a method for project management (such as *PMBOK® Guide* or PRINCE2®), (integrated) project management tools (planning, budgeting, and tracking) and supporting organization structures such as a project office. The sustainability aspects should be incorporated within these working methods:

Step 1: Understand the company's vision and ambition.

Changes to current working methods should fit and be based on the ambition and strategy of the company or division with respect to sustainability. The PMO leader therefore needs to have a clear understanding of this. The first step involves building your understanding of the company's vision with regards to sustainability. This might involve:

- desk research into the company's official documents;
- intensive dialogue with (top)-management, chief strategists, chief CSR officers and informal opinion leaders within the company to understand the real vision and ambition behind the formal documents.

Step 2: Goal setting: Create a company-specific framework for sustainability in projects.

In Chapter 4 we provided an integrated framework for sustainability in projects and project management. The specific nature of the company (for example, the market, the products, the production process) can imply that not all aspects of the framework are necessary for integrating sustainability in the projects of that company. Using the information from step 1, the project management sustainability maturity model will be the basis for the PMO leader to create a company-specific framework for their projects. Specific activities are:

- pinpoint the ambition (the desired situation) for the various elements of the maturity model. This gives specific indications of the relevant aspects and the appropriate depth of vision;
- discuss and communicate this ambition document with relevant stakeholders to verify that the framework represents and operationalizes the sustainability thoughts and ambition of the company;
- create new standards of project management products such as the business case, the project plan, project progress reports and the project management process.

Step 3: Analyze the current situation

In order to create a development plan you need to find out how and to what extent sustainability is already incorporated in the projects:

- assess and analyze various projects with the maturity model;
- determine in general terms the level of sustainability in projects and project management.

Step 4: Develop an implementation plan

Finally, the proposed and necessary changes need to be implemented. There will be changes with respect to the working methods (for example, the project management method) and changes in individual projects. Specific activities are:

- document and formalize the new working methods;
- develop the necessary products like the business case, progress reports and governance structures;
- train/educate the project managers in the new framework;
- determine for each individual project the necessary changes to comply with the reference framework;
- develop a process to assess the application of the framework within each (new) project.

6.2.2   Deliver skilled project managers

A second responsibility of the PMO leader is to deliver skilled project managers to the projects within the organization. The key message in this book is that the professionalism of project managers should drive responsibility for sustainability. Taking responsibility involves the desire to take responsibility as well as the knowledge and skills to do so; implying training and education. The PMO leader is responsible for providing this knowledge. The corresponding activities for getting started are:

Step 1: Provide awareness workshops

The workshops involve awareness raising among the project managers about responsibility for sustainability in projects and project management. In this workshop the following topics should be addressed:

- What is sustainability and why is it important for your business (Chapters 2 and 3)?
- Why are projects relevant in the context of sustainability (Chapter 3)?
- Why must project managers take responsibility?

- Where can project managers take responsibility in order to influence project results (section 4.1)?
- Overview of the maturity model (section 4.2) to show that specific actions are possible.

Step 2: Create individual and organizational Human Resources (HR) development plans

You need to develop an HR plan covering knowledge, skills and attitude in the context of sustainability for each individual project manager.

Step 3: Train and educate project managers

As a last step the PMO leader needs to provide the necessary training and education to the project managers. This should consist of:

- definition of sustainability;
- understanding the six principles of sustainability;
- the project management sustainability maturity model;
- the framework created for the company accompanied with the underlying arguments;
- applying the model to a specific project.

Step 4: Specify requirements to suppliers of project managers

On some occasions, project managers are hired from specialized project management companies. The PMO leader should define requirements with respect to sustainability (for example, skills, attitudes, knowledge) towards these suppliers.

6.2.3   Create an effective project management environment

The PMO leader plays an important role in developing an effective environment for project management within the organization. This is true in general but integrating sustainability into projects and project management requires various changes within the organization and runs into a more complex world as shown in the previous chapters.

The following actions need to take place:

Step 1: Embedding project management products into the formal governance of the organization

All project management products, such as the business case, project plan and progress reports, need formal embedding into the organization to create a steady direction and new way of governing projects. This formal situation is also necessary to assure the desired outcomes and prevent a fall-back towards regular governance practices.

Step 2: Positioning the project managers

The consulting skills of project managers will become a more critical success factor in taking responsibility and realizing sustainable results. Alongside this it is also necessary to create a formal position for (senior) project managers and the PMO leader to advise senior and top-level management on specific opportunities in realizing sustainability goals. Positioning the project managers in this way gives them the opportunity to realize the connection of Figure 5.2: feedback from the project to the strategy formulating processes of the organization.

This can be done by creating a formal position for a chief project officer who actively participates or assists the organization's portfolio manager and is intensive involved in discussing strategy and its implementation in general and through the definition of strategic projects.

6.3   GETTING STARTED AS A PROJECT SPONSOR

The project sponsor or project executive is interested in realizing the goal of the project which should include the sustainability goals. (In contrast with the senior user (in terms of PRINCE2®) who needs to work with the project results, which are the products of the project that need to be incorporated into the process/organization.) The start-up actions of the senior user are covered in the next section. The project sponsor should be very involved in the sustainability aspects; ultimately he or she is responsible for implementing and realizing the sustainability strategy of the company through the projects they are supervising. The project sponsor can direct and influence project results in two ways: firstly by directing a specific project and secondly by defining a portfolio of projects that realize a greater goal.

Step 1: Define ambition by project sponsor

It's important for the project sponsor to define the project or portfolio's ambition, from the perspective of the company´s vision and to specify critical focus areas in terms of sustainability. The sponsor can start by:

- formulating and prioritizing the relative order of the six principles of sustainability. This will reflect the importance and focus areas for implementing sustainability;

- selecting and specifying goals from the project management checklist (as presented in Chapter 3) in order to make sustainability as practical as possible;
- specifying the depth of vision the project sponsor wants to use as a viewpoint on the project.

Step 2a: Realize a bigger goal: Define the project portfolio from the ambition

In order to realize the overall sustainability goal of the organization or department, the project sponsor should define a portfolio of projects. The ambition, defined in step 1, will guide the direction of the projects.

Step 2b: Realize a bigger goal: Track ambition and direct it through the portfolio reporting process

Of course, having defined the portfolio, progress should be monitored and directed. A portfolio is not static, but dynamic, as it evolves, progresses and changes.

Step 3a: Direct a project: Start dialogue with project participants

During the definition process the project sponsor should start a dialogue with the project manager and/or the manager project management and the senior user. The dialogue is about the ambition of the project, and should give them a thorough understanding and identify the opportunities for realizing it.

Step 3b: Direct a project: Track ambition and requirements through the project reporting process

Following this definition process, the goals are tracked through to the reporting process. The project sponsor verifies that the corresponding project report reflects the agreements.

6.4   GETTING STARTED AS A PROJECT USER

The project user, for example the (team) manager, directly managing a production team, operates with processes, people, materials, tools and ICT. The project user needs to work with the project's deliverables and integrate them into the organization. Therefore the user may be less interested in the process of producing the deliverables.

Step 1: Obtain sustainability directions from the project sponsor

The starting point for the project user is a dialogue with the project sponsor around the sustainability directions behind specific goals and ambition.

Step 2: Analyze whether the project currently fulfils the sustainability directions

Thereafter, a dialogue with the project manager should take place to discuss whether the project fulfils the sustainability ambition and goals at that moment. This should be examined with respect to the elements and depth of vision of the project management sustainability maturity framework. You should discuss and document requisite changes and their impact.

Step 3: Make an implementation plan to realize the necessary changes

The project user can now alter or define the criteria for the project products in such a way that it reflects the ambition. The implementation plan should be discussed with the project sponsor and the project manager.

6.5   GETTING STARTED AS A PROJECT PORTFOLIO MANAGER

The project portfolio manager is responsible for the governing the portfolio of projects by which the organization realizes it strategic goals. Therefore he or she is responsible for the continuous alignment of strategic goals and the configuration of the project portfolio. Incorporating sustainability will change the evaluation of progress and alignment with strategic orientation. The execution of the portfolio management processes themselves will remain the same. Like the project manager, the portfolio manager can and should take a proactive approach towards realizing sustainability goals by challenging business management in their ambition and goal setting.

Step 1: Understand the company

First it is important to understand the company's overall mission and ambition with respect to sustainability. This overall view will set the sustainability goals and boundaries for other departments and the projects which run. The portfolio manager starts collecting information from the company by checking their website(s), annual reports and corporate social report (CSR-report) looking for sustainability statements. Interviews and discussions with senior management and staff will also provide a clear view on true sustainability goals.

Step 2: Formulate new portfolio evaluation criteria

From understanding the strategic orientation of the company, the portfolio manager should construct a relevant evaluation framework with respect to sustainability usable for defining and evaluation of projects and the portfolio in total:

- formulating and prioritizing the relative order of the six principles of sustainability. This will reflect the importance and focus areas for implementing sustainability;
- selecting and specifying goals from the project management checklist (as presented in Chapter 3) in order to make sustainability as practical as possible;
- specifying the depth of vision the organization wants to use as a viewpoint on the portfolio;
- discussing the evaluation criteria with senior management, senior staff (such as the PMO leader) in order to create broad acceptance.

Step 3: Create an implementation plan to apply these evaluation criteria to methodologies of portfolio management and organizational processes

In general, the processes of portfolio management will not differ, but the content (and complexity of decision making) will. So these new insights should be incorporated into the 'standard' portfolio processes and also other regular organization processes (such as the financial process).

Step 3a: Adjust portfolio management-related processes.

The content of the following portfolio processes should be altered, documented and communicated to the relevant stakeholders of these processes:

- stakeholder management;
- risk management;
- organizational governance;
- resource management.

Step 3b: Organization-related processes

The following portfolio management processes will also impact organizational processes. Changes to these processes should be discussed, documented and formally approved by the responsible persons for these processes:

- benefits management;
- financial management.

Step 4: Embedding project management products into the formal governance of the organization

All project management products, such as the business case, project plan and progress reports, need formal embedding into the organization and should be aligned with the portfolio evaluation criteria to create a steady direction and new way of governing projects. This formal situation is also necessary to assure the desired outcomes and prevent a fall-back towards regular governance practices.

Step 5: Assisting the PMO leader by positioning the project managers

The consulting skills of project managers will become a more critical success factor in taking responsibility and realizing sustainable results. Alongside this it is also necessary to create a formal position for (senior) project managers and the Manager project management to advise senior and top-level management on specific opportunities in realizing sustainability goals.

Positioning the project managers in this way, where they can contribute to the construction of the portfolio, gives them the opportunity to realize the connection of Figure 5.2: feedback from the project to the strategy-formulating processes of the organization.

This can be done by creating a formal position for a chief project officer who actively participates or assists the organization's portfolio manager and is actively involved in discussing strategy and its implementation in general and through the definition of strategic projects.

6.6   AND NOW... GET STARTED!

The core message of this book is that project managers should and can take responsibility for sustainability in their projects. Earlier chapters showed that they can influence the sustainability aspects of their projects and that their professionalism requires that they do so. This final chapter provided specific, specific actions that can be taken by project managers and other roles related to projects and project management. So project managers should take responsibility, can take responsibility and need to know how to take responsibility. And now... get started!

CORE SUBJECTS AND ISSUES OF SOCIAL RESPONSIBILITY FROM ISO 26000

The following table provides an overview of what ISO 26000 considers to be the main areas of interest for companies who aspire to be more sustainable. It summarizes seven social responsibility 'core subjects'. These core subjects are further broken down into 'issues', specific themes or activities a company should work on in order to contribute to sustainable development.

Table A.1   Core subjects and issues of ISO 26000

Core subject:	Issues:
Organizational governance	*No specific issues mentioned*
Human rights	Due diligence
	Human rights risk situations
	Avoidance of complicity
	Resolving grievances
	Discrimination and vulnerable groups
	Civil and political rights
	Economic, social and cultural rights
	Fundamental principles and rights at work
Labour practices	Employment and employment relationships
	Conditions of work and social protection
	Social dialogue
	Health and safety at work
	Human development and training in the workplace
The environment	Prevention of pollution
	Sustainable resource use

Table A.1   Core subjects and issues of ISO 26000 *concluded*

Core subject:	Issues:
	Climate change mitigation and adaptation
	Protection of the environment, biodiversity and restoration of natural habitats
Fair operating practices	Anti-corruption
	Responsible political involvement
	Fair competition
	Promoting social responsibility in the value chain
	Respect for property rights
Consumer issues	Fair marketing, factual and unbiased information and fair contractual practices
	Protecting consumers' health and safety
	Sustainable consumption
	Consumer service, support, and complaint and dispute resolution
	Consumer data protection and privacy
	Access to essential services
	Education and awareness
Community involvement and development	Community involvement
	Education and culture
	Employment creation and skills development
	Technology development and access
	Wealth and income creation
	Health
	Social investment

Source: Adapted from: http://www.iso.org/iso/discovering_iso_26000.pdf.

BOULDING'S CLASSIFICATION OF SYSTEMS

Level 1 Frameworks

The geography and anatomy of the universe: the patterns of electrons around a nucleus, the pattern of atoms in a molecular formula, the arrangement of atoms in a crystal, the anatomy of the gene, the mapping of the earth, and so on.

Level 2 Clockworks

The solar system or simple machines such as the lever and the pulley, even quite complicated machines like steam engines and dynamos fall mostly under this category.

Level 3 Thermostats

Control Mechanisms or Cybernetic Systems: the system will move to the maintenance of any given equilibrium, within limits.

Level 4 Cells

Open systems or self-maintaining structures. This is the level at which life begins to differentiate itself from not life.

Level 5 Plants

The outstanding characteristics of these systems (studied by the botanists) are first, a division of labour with differentiated and mutually dependent parts (roots, leaves, seeds, and so on), and second, a sharp differentiation between the genotype and the phenotype, associated with the phenomenon of equifinal or 'blueprinted' growth.

Level 6 Animals

Level characterized by increased mobility, teleological behaviour and self-awareness, with the development of specialized 'information receptors (eyes, ears, and so on) leading to an enormous increase in the intake of information.

Level 7 Human Beings

In, addition to all, or nearly all, of the characteristics of animal systems man possesses self consciousness, which is something different from mere awareness.

Level 8 Social Organizations

The unit of such systems is not perhaps the person but the 'role': that part of the person which is concerned with the organization or situation in question. Social organizations might be defined as a set of roles tied together with channels of communication.

Level 9 Trascendental Systems

The ultimates and absolutes and the inescapable unknowables that also exhibit systematic structure and relationship.

QUESTIONNAIRE OF THE SUSTAINABLE PROJECT MANAGEMENT MATURITY MODEL

The questionnaire consists of four sections and in total 31 questions. The first three sections cover descriptive questions regarding the respondent, the project that is assessed and the organizational context of the project. The fourth section consists of the actual assessment questions.

SECTION I. QUESTIONS REGARDING THE RESPONDENT
(4 questions)

What is your gender?
Select only one answer

A. [ ] Male
B. [ ] Female

What is your age?
Select only one answer

A. [ ] <25 years
B. [ ] 25–34 years
C. [ ] 35–44 years
D. [ ] 45–54 years
E. [ ] 55–64 years
F. [ ] 65 years or over

In which domain is your position?
Select only one answer

A. [ ] General Management
B. [ ] Commercial Management
C. [ ] Financial Management
D. [ ] IT Management
E. [ ] Project, Programme or Portfolio Management

F.   [ ]   Business Development
G.   [ ]   Consulting
H.   [ ]   Training or Education
I.   [ ]   Other

What level would you say your position is?
Select only one answer

A.   [ ]   Strategic Management
B.   [ ]   Tactical Management
C.   [ ]   Operational Management
D.   [ ]   Non-Management
E.   [ ]   Other

SECTION II. QUESTIONS REGARDING THE PROJECT
(6 questions)

What type of project is assessed?
Select only one answer

A.   [ ]   Building & Construction Public Infrastructure
B.   [ ]   Building & Construction Real Estate
C.   [ ]   Building & Construction Development
D.   [ ]   Organizational Change
E.   [ ]   Information Technology
F.   [ ]   Research and Development
G.   [ ]   Other

In what industry sector does the project take place?
Select only one answer

A.   [ ]   Agriculture
B.   [ ]   Industry
C.   [ ]   Energy
D.   [ ]   Building and Construction
E.   [ ]   Healthcare
F.   [ ]   Wholesale and Retail
G.   [ ]   Logistic Services
H.   [ ]   Financial Services
I.   [ ]   Facility and Real Estate Services
J.   [ ]   Legal Services
K.   [ ]   HR Services
L.   [ ]   ICT and Communication Services

M.   [ ]   Consulting
N.   [ ]   Public Administration
O.   [ ]   Education and Training
P.   [ ]   Other

Is the project international?
Select only one answer

A.   [ ]   No
B.   [ ]   Yes, the result of the project has international aspects
C.   [ ]   Yes, the resources working on the project are international
D.   [ ]   Yes, the suppliers in the project are international

In which geographical regions will the project have an impact?
Multiple answers allowed

A.   [ ]   Europe
B.   [ ]   North America
C.   [ ]   Central and/or South America
D.   [ ]   Asia
E.   [ ]   Africa
F.   [ ]   Australia

What is the approximate size of the project budget?
Select only one answer

A.   [ ]   < 1 million €
B.   [ ]   Between 1 and 10 million €
C.   [ ]   Between 10 and 100 million €
D.   [ ]   > 100 million €

How many business partners (suppliers, subcontractors, etc.) will participate in the project?
Select only one answer

A.   [ ]   0
B.   [ ]   1–5
C.   [ ]   6–15
D.   [ ]   16–50
E.   [ ]   > 50

SECTION III. QUESTIONS REGARDING THE ORGANIZATIONAL CONTEXT OF THE PROJECT
(2 questions)

What is the position of sustainability in the strategy of the organization that commissions the project?
Multiple answers allowed. Please tick all answers that are applicable.

A.   [ ]   The strategy of the organization does not include any statements or ambitions regarding sustainability.

B.   [ ]   The strategy of the organization mentions a wise use of natural resources and/or social responsibility as one of the guiding principles for the selection of resources of the organization.

C.   [ ]   The strategy of the organization mentions a wise use of natural resources and/or social responsibility as one of the guiding principles for the (design of the) business processes of the organization.

D.   [ ]   The strategy of the organization mentions a wise use of natural resources and/or social responsibility as one of the guiding principles for the (design of the) business model of the organization.

E.   [ ]   The strategy of the organization mentions a wise use of natural resources and/or social responsibility as one of the guiding principles for the (development of) products and services of the organization.

Does the organization that commissions the project have any form of sustainability reporting?
Multiple answers allowed. Please tick all answers that are applicable.

A.   [ ]   No, the organization does not have any specific form of sustainability reporting.

B.   [ ]   Yes, the organization reports on their contribution as a part or section of the regular company reports (e.g. the Annual Report).

C.   [ ]   Yes, the organization reports on their contribution as a separate periodic sustainability report in a self-developed format.

D.   [ ]   Yes, the organization reports on their contribution as a separate periodic sustainability report in a format that is based on the sustainability reporting guidelines of the GRI.

SECTION IV. QUESTIONS REGARDING THE ASSESSMENT OF SUSTAINABILITY ASPECTS IN THE PROJECT

In all questions, multiple answers are allowed. Please tick all answers that are applicable for the project.

Please answer all questions twice. The first time for the actual situation. The second time for the 'desired' situation as you would prefer it.

Profit perspective
(4 questions)

Direct (Financial) Benefits
Which types of benefits are recognized in the business case of the project?
Multiple answers allowed. Please tick all answers that are applicable.

	Actual situation	Desired situation	
A.	[ ]	[ ]	Benefits are not explicitly recognized or no business cases are made for the project.
B.	[ ]	[ ]	Benefits are recognized in terms of cost savings or reduced use of resources.
C.	[ ]	[ ]	Benefits are recognized in terms of improved business processes.
D.	[ ]	[ ]	Benefits are recognized in terms of extra revenues from new business models for existing products and services.
E.	[ ]	[ ]	Benefits are recognized in terms of extra revenues from innovated products or services.

Managerial Flexibility and Optionality
To what extent does the project allow for future decision making and real options?
Multiple answers allowed. Please tick all answers that are applicable.

	Actual situation	Desired situation	
A.	[ ]	[ ]	Projects are designed as optimal as possible, given our current knowledge. Future decisions may or may not be included in this design, but are not a design criterion as such.

B.   [ ]   [ ]   Projects are designed as optimal as possible, and decisions regarding materials, resources and suppliers are made as late as possible to allow for flexibility in the execution of the project.

C.   [ ]   [ ]   Projects are designed as optimal as possible, and decisions regarding project (production) processes and outsourcing partners are made as late as possible to allow for flexibility in the execution of the project.

D.   [ ]   [ ]   Projects are designed as optimal as possible, and the exact requirements of the project deliverables are made as late as possible to allow for flexibility in the execution of the project.

E.   [ ]   [ ]   Projects are designed as optimal as possible, and the exact requirements of the project goal, result and deliverables are made as late as possible to allow for flexibility in the execution of the project.

Project Reporting
Which items are reflected in the project's (progress) reports?
Multiple answers allowed. Please tick all answers that are applicable.

	Actual situation	Desired situation	
A.	[ ]	[ ]	The project does not formally report progress.
B.	[ ]	[ ]	The project (progress) reports show items such as activities commenced, activities completed, budget spent, budget still required, total budget, issues and risks, all in terms of 'plan' and 'actual'.
C.	[ ]	[ ]	The project (progress) reports show also lessons learned and improvements to the project.
D.	[ ]	[ ]	The project (progress) reports show also suggestions to (radically) change the way the project is being designed and delivered.
E.	[ ]	[ ]	The project (progress) reports show also changes (e.g. market conditions) that may have an effect on the value and business case of the project's result.

Investment Evaluation

Which evaluation methods are used in the selection of projects?
Multiple answers allowed. Please tick all answers that are applicable.

	Actual situation	Desired situation	
A.	[ ]	[ ]	No formal selection methods are used, projects are selected based on the availability of funds to invest.
B.	[ ]	[ ]	Projects are evaluated and selected predominantly based on the pay-back period of the investment.
C.	[ ]	[ ]	Projects are evaluated and selected predominantly based on the return on investment or net present value of the investment.
D.	[ ]	[ ]	Projects are evaluated and selected predominantly based on their long term strategic value, in combination with their short/medium term returns.
E.	[ ]	[ ]	Projects are evaluated and selected predominantly based on a balanced set of quantitative and qualitative criteria that reflect both long term and short term perspectives and also economical, social and environmental aspects.

Planet Perspective

(7 questions)

Procurement

Based on which criteria are suppliers for the project selected?
Multiple answers allowed. Please tick all answers that are applicable.

	Actual situation	Desired situation	
A.	[ ]	[ ]	Suppliers for the project are selected based on price.
B.	[ ]	[ ]	Suppliers for the project are selected based on location for minimizing transport.
C.	[ ]	[ ]	Suppliers for the project are selected based on their own use of natural resources and policies to enhance environmental sustainability.

D.	[ ]	[ ]	Suppliers for the project are selected based on how their know-how and partnership helps our project to be delivered in a more sustainable way.
E.	[ ]	[ ]	Suppliers for the project are selected based on how their know-how and partnership helps our product and services to aid sustainability.

Materials

Based on which criteria are materials for the project selected?
Multiple answers allowed. Please tick all answers that are applicable.

	Actual situation	Desired situation	
A.	[ ]	[ ]	Materials for the project are selected based on technical and functional requirements and their costs.
B.	[ ]	[ ]	Materials for the project are also selected based on the waste they cause in and for the project.
C.	[ ]	[ ]	Materials for the project are also selected based on the energy consumption and/or pollution incorporated in the materials because of their production process.
D.	[ ]	[ ]	Materials for the project are also selected based on the energy consumption and/or pollution incorporated in the materials because of their production and logistics processes.
E.	[ ]	[ ]	Materials for the project are also selected based on their reuse capabilities and value.

Energy

Does the project have any specific policies regarding its energy consumption?
Multiple answers allowed. Please tick all answers that are applicable.

	Actual situation	Desired situation	
A.	[ ]	[ ]	Next to general policies on energy consumption within the organization, the project does not have specific policies and the energy consumption.
B.	[ ]	[ ]	There are policies in the project to promote the smart use of energy and where possible, energy saving equipment is used.

C.   [ ]   [ ]   Where possible, energy consumption is actively kept to a minimum and the necessary energy used is acquired as 'green' energy.

D.   [ ]   [ ]   Minimizing energy consumption is one of the parameters in the design of the project delivery processes.

E.   [ ]   [ ]   Minimizing energy consumption is one of the parameters in the design of the project deliverable and result.

Water

Does the project have any specific policies regarding its water consumption and pollution?

Multiple answers allowed. Please tick all answers that are applicable.

	Actual situation	Desired situation	
A.	[ ]	[ ]	Next to general policies on water consumption and pollution within the organization, the project does not have specific policies.
B.	[ ]	[ ]	There are policies in the project to promote the smart use of water and where possible, water saving equipment is used.
C.	[ ]	[ ]	Water consumption is actively kept to a minimum and where possible, the project recycles its water and/or purifies it.
D.	[ ]	[ ]	Minimizing water consumption and pollution is one of the parameters in the design of the project delivery processes. The necessary water used is recycled and/or purified before disposal.
E.	[ ]	[ ]	Minimizing water consumption and pollution is one of the parameters in the design of the project deliverable and result. The project result actively minimizes water consumption and pollution and the necessary water used is recycled and/or purified before disposal.

Waste

In which way does the project try to minimize its waste?
Multiple answers allowed. Please tick all answers that are applicable.

	Actual situation	Desired situation	
A.	[ ]	[ ]	No specific policies on this point.
B.	[ ]	[ ]	Waste in the project is separated in recyclable and non-recyclable and collected by the local waste handling companies.
C.	[ ]	[ ]	The project has policies (e.g. double sided printing) to minimize waste.
D.	[ ]	[ ]	The project delivery processes are designed to minimize waste and necessary waste is as much as possible recycled in the project itself.
E.	[ ]	[ ]	The project deliverable and result are designed to minimize waste and necessary waste is as much as possible recycled in the deliverable itself.

Travel

To what extent does the project apply travel policies that consider environmental aspects?
Multiple answers allowed. Please tick all answers that are applicable.

	Actual situation	Desired situation	
A.	[ ]	[ ]	Travelling in the project is based on necessity for the project's activities and deliverables. Means of travel are selected on costs and time.
B.	[ ]	[ ]	Travelling in the project is based on necessity for the project's activities and deliverables, but means of travel are selected considering environmental aspects.
C.	[ ]	[ ]	Travelling in the project is based on necessity and minimized by actively promoting and facilitating the use of alternatives for travelling (e.g. video conferencing).
D.	[ ]	[ ]	The project delivery processes are designed to minimize travelling in the project.

E.   [ ]   [ ]   The project deliverable and result are designed to minimize travelling.

Project Reporting
Does the project's (progress) reports reflect indicators of environmental sustainability?
Multiple answers allowed. Please tick all answers that are applicable.

	Actual situation	Desired situation	
A.	[ ]	[ ]	The project does not formally report progress.
B.	[ ]	[ ]	The project (progress) reports reflect indicators of environmental sustainability with respect to used (physical) resources.
C.	[ ]	[ ]	The project (progress) reports reflect indicators of environmental sustainability with respect to the project delivery process.
D.	[ ]	[ ]	The project (progress) reports reflect indicators of environmental sustainability with respect to the project deliverable or result.
E.	[ ]	[ ]	The project (progress) reports reflect indicators of environmental sustainability with respect to the use and disposal of the project deliverable or result.

People Perspective
(8 questions)

Labour Practices and Decent Work
To what extent does the project apply policies or standards for labour practices and decent work?
Multiple answers allowed. Please tick all answers that are applicable.

	Actual situation	Desired situation	
A.	[ ]	[ ]	The project complies with applicable standards and regulations on labour practices or decent work.
B.	[ ]	[ ]	The project also requires its suppliers and partners to practise good labour practices and decent work.

C.   [ ]   [ ]   The project actively (re) designs its project delivery
                       processes in a way that labour practices are improved
                       and/or on a high level.

D.   [ ]   [ ]   The project's deliverable and result is designed to improve
                       labour practices and decent work in the organization that
                       commissioned the project.

E.   [ ]   [ ]   The project's deliverable and result is designed to improve
                       labour practices and decent work in the community in
                       which the result is used or aimed at.

Health and Safety
To what extent does the project apply policies or standards for health and safety?
Multiple answers allowed. Please tick all answers that are applicable.

	Actual situation	Desired situation	
A.	[ ]	[ ]	The project complies with applicable standards and regulations on health and safety.
B.	[ ]	[ ]	The project also requires its suppliers and partners to practise good health and safety practices.
C.	[ ]	[ ]	The project actively (re) designs its project delivery processes in a way that health and safety risks are minimized.
D.	[ ]	[ ]	The project's deliverable and result is designed to improve health and safety conditions in the organization that commissioned the project.
E.	[ ]	[ ]	The project's deliverable and result is designed to improve health and safety conditions in the community in which the result is used or aimed at.

Training, Education and Organizational Learning
To what extent does the project includes training, education and development of stakeholders?
Multiple answers allowed. Please tick all answers that are applicable.

	Actual situation	Desired situation	
A.	[ ]	[ ]	The project includes activities for training and education of end users as part of the project's deliverables (if applicable).

B.   [ ]   [ ]   The project includes activities for training and education of team members for improved individual and team performance in the project.

C.   [ ]   [ ]   The project includes activities for training and education of team members and partners for improved individual and team performance after the project has finished.

D.   [ ]   [ ]   The project includes activities for developing the (project) competences of all stakeholders involved.

E.   [ ]   [ ]   The project's result involves or includes activities for the development of the community that is effected by the project or the project result.

Diversity and Equal Opportunity

To what extent does the project apply policies or standards for diversity and equal opportunity that reflects the society it operates in?
Multiple answers allowed. Please tick all answers that are applicable.

	Actual situation	Desired situation	
A.	[ ]	[ ]	The project complies with applicable standards and regulations on equal opportunity in terms of gender, race, religion, etc.
B.	[ ]	[ ]	The project also requires its suppliers and partners to practise diversity practices and provide equal opportunity in terms of gender, race, religion, etc.
C.	[ ]	[ ]	The project actively (re) designs its project delivery processes in a way (e.g. by designing part-time jobs) that diversity and equal opportunity are promoted and stimulated.
D.	[ ]	[ ]	The project's deliverable and result are designed to improve diversity and equal opportunity in the organization that commissioned the project.
E.	[ ]	[ ]	The project's deliverable and result are designed to improve diversity and equal opportunity in the community in which the result is used or aimed at.

Human Rights

To what extent does the project apply policies or standards for respecting and improving human rights like non-discrimination, freedom of association and no child labour?

Multiple answers allowed. Please tick all answers that are applicable.

	Actual situation	Desired situation	
A.	[ ]	[ ]	The project complies with applicable standards and regulations on human rights and stimulates improvement of these rights where applicable.
B.	[ ]	[ ]	The project also requires its suppliers and partners to respect and improve human rights where possible.
C.	[ ]	[ ]	The project actively (re) designs its project delivery processes in a way that human rights are improved and/or on a high level.
D.	[ ]	[ ]	The project's deliverable and result are designed to respect and improve human rights in the organization that commissioned the project.
E.	[ ]	[ ]	The project's deliverable and result are designed to respect and improve human rights in the community in which the result is used or aimed at.

Society and Customers

To what extent does the project take a social responsibility towards the society it operates in?

Multiple answers allowed. Please tick all answers that are applicable.

	Actual situation	Desired situation	
A.	[ ]	[ ]	For the general acceptance of the project and its results, the project recognizes a social responsibility towards the external stakeholders in the society it operates in.
B.	[ ]	[ ]	The project also requires its suppliers and partners to take on social responsibility towards the external stakeholders in the society they operates in.

C. [ ] [ ] The project actively (re) designs its project delivery processes in a way that translates its social responsibility towards the external stakeholders in the society it operates in.

D. [ ] [ ] The project's deliverable and result are designed in a way that translates its social responsibility towards the external stakeholders in the society it operates in.

E. [ ] [ ] The project's deliverable and result are designed in a way that translates its social responsibility towards the total society.

Bribery and Anti-Competitive Behaviour

To what extent does the project reject bribery and anti-competitive behaviour?
Multiple answers allowed. Please tick all answers that are applicable.

	Actual situation	Desired situation	
A.	[ ]	[ ]	The project rejects bribery and anti-competitive behaviour and holds responsible team members accountable.
B.	[ ]	[ ]	The project also requires its suppliers and partners to reject bribery and anti-competitive behaviour.
C.	[ ]	[ ]	The project actively (re) designs its project delivery processes and trains its project members to prevent bribery and anti-competitive behaviour.
D.	[ ]	[ ]	The project actively (re) designs its project deliverable and results in a way that bribery and anti-competitive behaviour is prevented in the organization that commissioned the project.
E.	[ ]	[ ]	The project actively (re) designs its project deliverable and results in a way that bribery and anti-competitive behaviour is prevented in the community in which the result is used or aimed at.

Project Reporting

Does the project's (progress) reports reflect indicators of social sustainability?
Multiple answers allowed. Please tick all answers that are applicable.

	Actual situation	Desired situation	
A.	[ ]	[ ]	The project does not formally report progress.
B.	[ ]	[ ]	The project (progress) reports reflect indicators of social sustainability with respect to used (physical) resources.
C.	[ ]	[ ]	The project (progress) reports reflect indicators of social sustainability with respect to the project delivery process.
D.	[ ]	[ ]	The project (progress) reports reflect indicators of social sustainability with respect to the project deliverable or result.
E.	[ ]	[ ]	The project (progress) reports reflect indicators of social sustainability with respect to the use and disposal of the project deliverable or result.

DESCRIPTION OF THE ICB® VERSION 3.0 COMPETENCES

Table D.1   **Description of the ICB® Version 3.0 competences**

Technical competences	
Competence	*Brief description*
1.01 Project management success	The project manager recognizes and appreciates the criteria and conditions of project success in the eyes of the interested parties.
1.02 Interested parties	The project manager recognizes and identifies the different interested parties in the project. (Note: 'interested parties' is used as synonym with stakeholders.)
1.03 Project requirements & objectives	The project manager recognizes and understands the goals, requirements and conditions of the project.
1.04 Risk & opportunity	The project manager recognizes and understands the risks of the project and manages these adequately.
1.05 Quality	The project manager understands the quality aspects of both project result as project execution and manages the realization of these aspects.
1.06 Project organization	The project manager designs, establishes and maintains an efficient and effective division of tasks in appropriate roles, responsibilities and capabilities for the project.
1.07 Teamwork	The project manager recognizes the distinct qualities of the different team members and moulds them into an effective team.
1.08 Problem resolution	The project manager identifies (potential) problems in an early stage and is capable of solving the issues at hand.
1.09 Project structures	The project manager organizes the project team and its relations with stakeholders in effective organizational and communication structures.

Table D.1   Description of the ICB® Version 3.0 competences *continued*

Technical competences	
Competence	*Brief description*
1.10 Scope & deliverables	The project manager specifies the project objective and assignment in specific project results, activities and work packages and understands how these are interrelated.
1.11 Time & project phases	Understanding the interrelations, the project manager plans and schedules the project activities and groups them into a clear project phasing.
1.12 Resources	The project manager identifies, recognizes and organizes the (personal as well as material) resources required for the project.
1.13 Cost & finance	The project manager plans and manages the cash flows related to the project and acquires sufficient funding.
1.14 Procurement & contract	The project manager qualifies, selects and contracts suppliers to the project, plans the purchases and coordinates the deliveries.
1.15 Changes	The project manager handles requests for change efficiently and effectively taking into account the scope of the project and the impact of the changing requirements.
1.16 Control & reports	The project manager directs the realization of the project plan, monitors the progress of activities, reports project progress and anticipates contingencies.
1.17 Information & documentation	The project manager plans, collects, archives and analyzes the project documentation and information.
1.18 Communication	The project manager is skilled in communication and deploys his skills efficiently and effectively. He is also perceptive of verbal and non-verbal communication of others.
1.19 Start-up	The project manager realizes an adequate project start-up that creates commitment of team members and interested parties for the project goal and plan.
1.20 Close-out	The project manager realizes an adequate project closure that transfers the results of the project to the project owner and dismisses the project organization from their duties.

Table D.1   Description of the ICB® Version 3.0 competences *continued*

Behavioural competences	
Competence	*Brief description*
2.01 Leadership	The project manager stimulates and motivates team members and interested parties to act in the interest of the project and show efficient and effective behaviour.
2.02 Engagement & motivation	The project manager is personally committed to and motivated for the project.
2.03 Self-control	The project manager organizes his job effectively and efficiently and dismisses unnecessary tension or pressure.
2.04 Assertiveness	The project manager is adequately assertive and convincing to ensure successful project realization
2.05 Relaxation	The project manager de-escalates conflicts and tension and facilitates effective teamwork.
2.06 Openness	The project manager creates an open atmosphere within his project team that allows new team members to immediately feel at ease. He is also open to feedback and comments.
2.07 Creativity	The project manager explores problems and issues from different and unexplored angles and is able to develop new and innovative solutions.
2.08 Results orientation	The project manager does not lose his focus on the project goals and the interests of the interested parties and achieves project results.
2.09 Efficiency	The project manager utilizes project resources and team members efficiently and effectively.
2.10 Consultation	The project manager analyses issues and situations, seeks advice and new insights, weights pros and cons of different alternatives and makes informed decisions.
2.11 Negotiation	The project manager creates consensus and cooperation for his decisions.
2.12 Conflict & crisis	The project manager anticipates on or recognizes potential conflicts of interest or crises in an early stage and develops solutions that prevent or solve the issue.
2.13 Reliability	The project manager is reliable in his behaviour and does not harm the confidence put in him.
2.14 Values appreciation	The project manager recognizes the beliefs and values of team members and interested parties and respects these.
2.15 Ethics	The project manager understands ethic and moral values and acts accordingly.

Table D.1   Description of the ICB® Version 3.0 competences *concluded*

Contextual competences	
Competence	*Brief description*
3.01 Project orientation	The project manager understands the rationale for the project and is aware of the organizational context of the project.
3.02 Program orientation	The project manager is capable of aligning programme goals to business strategy and develops new proposals for new projects supporting this strategy.
3.03 Portfolio orientation	The project manager advises the organization about effective project and programme priorities and about the portfolio management process.
3.04 Project, program & portfolio orientation	The project manager creates awareness in the organization of the role of portfolios, programmes and projects in the realization of the organization's strategy.
3.05 Permanent organization	The project manager is aware of the complex relations between the project its surrounding organizations and is capable of managing these relations in an effective manner.
3.06 Business	The project manager has knowledge and understanding of the specific business and business processes of the project owner's organization.
3.07 Systems, products & technology	The project manager understands the causes of developments and the effects of actions in the project and is able to manage these relations effectively.
3.08 Personnel management	The project manager recruits, selects, develops, appraises and rewards his team members in a way that stimulates effective behaviour and successful teamwork.
3.09 Health, security, safety & environment	The project manager is aware of health, security, safety and environmental aspects of the project and manages these adequately.
3.10 Finance	The project manager has adequate knowledge of and insight in the financial and administrative processes of the project and integrates these aspects in his actions.
3.11 Legal	The project manager is aware of legal, compliancy and liability aspects of the project and manages these adequately.

DESCRIPTION OF THE *PMCD FRAMEWORK* COMPETENCES

The *Project Management Competence Development (PMCD) Framework* from the Project Management Institute (PMI) provides a framework for the definition, assessment and development of project manager competence. It defines key dimensions of competence and identifies those competences that are most likely to impact project managers' performance.

The project manager competence consists of three separate dimensions:

1. Knowledge competence: what the project manager knows about the application of processes, tools and techniques.
2. Performance competence: how the project manager applies project management knowledge to meet project requirements
3. Personal competence how the project manager behaves when performing activities in the project environment; their attitude and core personality characteristics.

The *PMCD Framework* is based upon the principles and processes of the *PMBOK®Guide*. It describes the generic competences needed in most projects in most companies in most industries. In some industries there may be technical skills that are particularly relevant to the industry or covered by specific domain, regulatory or legal requirements. The personal and performance competences in the *PMCD Framework* are comprised of units of competence. The knowledge competence is not detailed in within the *PMCD Framework*. The *PMCD Framework* breaks the desired competences into a simple structure. At the highest level there are Units of Competence which divides competences in major segments, typically representing a major function or activity. At the next tier there are elements of competence, which are the building blocks of each competence. They describe in output terms, actions or outcome that are demonstrable or accessible. Elements associated with performance competences are project outcomes or results. Elements associated with personal competences are project managers' behaviour descriptions. Each element is further defined by performance criteria that specify the actions required to demonstrate competent performance and by types of evidence (specific products that the action has been completed). Table E.1 only shows the competence elements. For the performance criteria and evidence we refer to the official publications on the *PMCD Framework* Second Edition.

Table E.1   Competence elements of the *PMCD Framework*

Performance competences	
Initiating a project	• Projects aligned with organizational objectives and customer needs • Preliminary scope statements reflects stakeholder needs and expectations • High-level risks, assumptions and constraints are understood • Stakeholders identified and their needs understood • Project charter approved
Planning a project	• Project scope agreed • Project schedule approved • Cost budget approved • Project team identified with roles and responsibilities • Communications activities agreed • Quality management process established • Risk reponse plan approved • Integrated change control processes defined • Procurement plan approved • Project plan approved
Executing a project	• Project scope achieved • Project stakeholders expectations managed • Human resources managed • Quality managed against plan • Material resources managed
Monitoring and controlling the project	• Project tracked and status communicated to stakeholders • Project change is managed • Quality is monitored and controlled • Risk is monitored and controlled • Project team is managed • Contracts administered
Closing a project	• Project outcomes accepted • Project resources released • Stakeholder perceptions measured and analyzed • Project formally closed
Personal competences	
Communicating	• Actively listens, understands and responds • Maintains lines of communication • Ensures quality of information • Tailors communication to audience

Table E.1   Competence elements of the *PMCD Framework concluded*

Personal competences	
Leading	• Creates a team environment that promotes high performance • Builds and maintains effective relations • Motivates and mentors project team members • Takes accountability for delivering the project • Uses influences skills when required
Managing	• Builds and maintains the project team • Plans and manages for project success in an organized manner • Resolves conflicts involving project team or stakeholder
Cognitive ability	• Take a holistic view of the project • Effectively resolves issues and solves problems • Uses appropriate project management tools and techniques • Seeks opportunities to improve project outcome
Effectiveness	• Resolves project problems • Maintains stakeholder involvement, motivation and support • Changes at the required pace to meet project needs • Uses assertiveness when required
Professionalism	• Demonstrates commitment to the project • Operates with integrity • Handles personal and team adversity in a suitable manner • Manages a diverse workforce • Resolves individual and organizational issues objectively

GRAVES' MODEL OF VALUE SYSTEMS

A value system is the image a person has on the world with convictions connected to them. Convictions about what is good or bad, important and not important. When life conditions change and it's necessary to think different, then people also change. Clare W. Graves was a professor of psychology and originator of a theory of adult human development. He defined eight values systems and argues that human nature is not fixed: humans are able, when forced by life conditions, to adapt to their environment by constructing new, more complex, conceptual models of the world that allow them to handle the new problems. Each new model includes and transcends all previous models. Within the model, individuals and cultures embody a mixture of the value patterns, with varying degrees of intensity in each. Attaining higher stages of development is not synonymous with attaining a 'better' or 'more correct' values system. All stages co-exist in both healthy and unhealthy states, whereby any stage of development can lead to undesirable outcomes with respect to the health of the human and social environment. The following value system levels are defined:

World Value System One – Beige – Instinct – The concerns here are for survival with a major emphasis on meeting basic physical needs.

World Value System Two – Purple – Tribe – Membership of a tribe or family or team is most important and people will repress their own needs for the benefit of the tribe.

World Value System Three – Red – Power – There is a shift back to the individual here and people coming from this value system only really recognize and accept power.

World Value System Four – Blue – Order – To regain the necessary balance, order must be re-established. Hierarchy and knowing the difference between right and wrong is of major importance.

World Value System Five – Orange – Technology – The status quo can be turgid and stop progress and so to become more efficient, corners must be cut and better ways found. Technology and science are the best ways to achieve profit in this world of unlimited resources.

World Value System Six – Green – Humanity – A concern for people comes back, as the previous value system can seem insensitive and heartless and we need to look after one another if we are to survive in this world of limited resources.

World Value System Seven – Yellow – Systems – For the first time, we can recognize the other systems for what they are and work with them. Up to this point each worldview believes that its view is the best view, the only view. Recognizing and using the systems will lead to more efficiency and utilizing fewer resources.

World Value System Eight – Turquoise – Global – There is now a recognition and utilization of all of the systems within systems in a world which is interconnected in ways that very few can see or understand.

PORTFOLIO, PROGRAMME & PROJECT MANAGEMENT MATURITY MODEL (P3M3®)

The Portfolio, Programme & Project Management Maturity Model (P3M3®) contains three models that enable independent assessment. There are no interdependencies between the models, so an organization may be better at programme management than it is at project management, for example. The models are:

- Portfolio Management (PfM3)
- Programme Management (PgM3)
- Project Management (PjM3).

Similar to the Capability Maturity Model (CMM) process maturity framework, P3M3® is described by a five level maturity framework:

- Level 1 – awareness of process
- Level 2 – repeatable process
- Level 3 – defined process
- Level 4 – managed process
- Level 5 – optimized process.

These levels comprise the structural components of P3M3®. Table G.1 provides a description of the application of the maturity levels to Project Management, Programme Management and Portfolio Management.

As organizations move up through the maturity levels, the quality and effectiveness of the processes and practices increase correspondingly. This incremental nature of process improvement is a key feature of P3M3®.

Table G.1   Maturity levels of the P3M3® framework

Project Management maturity levels	
Level 1 Awareness of process	The organization recognizes projects and organizes these separate from its ongoing business. These projects may be run with no standard process or tracking system.
Level 2 Repeatable process	The organization recognizes projects, organizes these separate from its ongoing business and has these projects run with specific processes and procedures, to a minimum specified standard. However, consistency or coordination between projects may be limited.
Level 3 Defined process	The organization recognizes projects, organizes these separate from its ongoing business and ensures that projects are run with (centrally) controlled specific processes and procedures. Individual projects may flex within these processes to suit the particular project.
Level 4 Managed process	The organization recognizes projects, organizes these separate from its ongoing business, ensures that these projects are run with (centrally) controlled specific processes and procedures and retains specific measurements on its project management performance in order to improve future project performance.
Level 5 Optimized process	The organization recognizes projects, organizes these separate from its ongoing business, ensures that these projects are run with (centrally) controlled specific processes and procedures and undertakes continuous process improvement with proactive problem and technology management for projects in order to improve its ability to depict performance over time and optimize processes.
Programme Management maturity levels	
Level 1 Awareness of process	The organization recognizes programmes and runs these differently from projects. These programmes may be run with no standard process or tracking system.
Level 2 Repeatable process	The organization recognizes programmes and runs these differently from projects and has these programmes run with specific processes and procedures, to a minimum specified standard. However, consistency or coordination between programmes may be limited.

Table G.1   Maturity levels of the P3M3® framework *continued*

Programme Management maturity levels	
Level 3 Defined process	The organization recognizes programmes and runs these differently from projects and ensures that programmes are run with centrally controlled specific processes and procedures. Individual programmes may flex within these processes to suit the particular programme.
Level 4 Managed process	The organization recognizes programmes and runs these differently from projects and ensures that programmes are run with centrally controlled specific processes and procedures and retains specific measurements on its programme management performance in order to improve future programme performance.
Level 5 Optimized process	The organization recognizes programmes and runs these differently from projects and ensures that programmes are run with centrally controlled specific processes and procedures and undertakes continuous process improvement with proactive problem and technology management for programmes in order to improve its ability to depict performance over time and optimize processes.
Portfolio Management maturity levels	
Level 1 Awareness of process	The organization's Executive Board recognizes programmes and projects, and maintains an (informal) list of its investments in programmes and projects. There may be no formal tracking and documenting process.
Level 2 Repeatable process	The organization's Executive Board recognizes programmes and projects, and ensures that these programmes and/or projects are run with specific processes and procedures, to a minimum specified standard. However, consistency or coordination may be limited.
Level 3 Defined process	The organization's Executive Board recognizes programmes and projects, and ensures that these programmes and/or projects are run with centrally controlled specific processes and procedures. Individual programmes and/or projects may flex within these processes to suit the particular programme/project. The organization also has a portfolio management process in place to coordinate and prioritize its investments in programmes and projects.

Table G.1   Maturity levels of the P3M3® framework *concluded*

Portfolio Management maturity levels	
Level 4 Managed process	The organization's Executive Board recognizes programmes and projects, and ensures that these programmes and/or projects are run with centrally controlled specific processes and procedures. Individual programmes and/or projects may flex within these processes to suit the particular programme/project. The organization also has a portfolio management process in place to coordinate and prioritize its investments in programmes and projects. It retains specific measurements on its programme and project management performance in order to improve the performance of its total portfolio of programmes and/or projects.
Level 5 Optimized process	The organization's Executive Board recognizes programmes and projects, and ensures that these programmes and/or projects are run with centrally controlled specific processes and procedures. Individual programmes and/or projects may flex within these processes to suit the particular programme/project. The organization also has a portfolio management process in place to coordinate and prioritize its investments in programmes and projects. It retains specific measurements on its programme and project management performance in order to improve the performance of its total portfolio of programmes and/ or projects, and undertakes continuous process improvement with proactive problem and technology management in order to improve its ability to depict performance over time and optimize processes.

P3M3® focuses on the following seven Process Perspectives defining the key characteristics of a mature organization. They exist in all three models and can be assessed at all five maturity levels:

- *Management control:*
  This covers the internal controls of the initiative and how its direction of travel is maintained throughout its life cycle, with appropriate break points to enable it to be stopped or redirected by a controlling body if necessary.
- *Benefits management:*
  Benefits management is the process that ensures that the desired business change outcomes have been clearly defined, are measurable and are ultimately realized through a structured approach and with full organizational ownership.

- *Financial management:*
  Finance is an essential resource that should be a key focus for initiating and controlling initiatives. Financial management ensures that the likely costs of the initiative are captured and evaluated within a formal business case and that costs are categorized and managed over the investment life cycle.
- *Stakeholder management:*
  Stakeholders are key to the success of any initiative.
- *Risk management:*
  This views the way in which the organization manages threats to, and opportunities presented by, the initiative.
- *Organizational governance:*
  This looks at how the delivery of initiatives is aligned to the strategic direction of the organization. It considers how start-up and closure controls are applied to initiatives and how alignment is maintained during an initiative's life cycle. This differs from management control, which views how control of initiatives is maintained internally, as this perspective looks at how external factors that impact on initiatives are controlled (where possible, or mitigated if not) and used to maximize the final result.
- *Resource management:*
  Resource management covers management of all types of resources required for delivery. These include human resources, buildings, equipment, supplies, information, tools and supporting teams. A key element of resource management is the process for acquiring resources and how supply chains are utilized to maximize effective use of resources.

Source: www.p3m3-officialsite.com

VARIABLES OF THE EFQM EXCELLENCE MODEL

LEADERSHIP

Excellent organizations have leaders who shape the future and make it happen, acting as role models for its values and ethics and inspiring trust at all times. They are flexible, enabling the organization to anticipate and react in a timely manner to ensure the ongoing success of the organization.

1. *Strategy*
   Excellent organizations implement their mission and vision by developing a stakeholder focused strategy. Policies, plans, objectives and processes are developed and deployed to deliver the strategy.
2. *People*
   Excellent organizations value their people and create a culture that allows the mutually beneficial achievement of organizational and personal goals. They develop the capabilities of their people and promote fairness and equality. They care for, communicate, reward and recognize, in a way that motivates people, builds commitment and enables them to use their skills and knowledge for the benefit of the organization.
3. *Partnerships and resources*
   Excellent organizations plan and manage external partnerships, suppliers and internal resources in order to support strategy and policies and the effective operation of processes.
4. *Processes, products and services*
   Excellent organizations design, manage and improve processes to generate increasing value for customers and other stakeholders.
5. *Customer results*
   Excellent organizations comprehensively measure and achieve outstanding results with respect to their customers.
6. *People results*
   Excellent organizations comprehensively measure and achieve outstanding results with respect to their people.

7. *Society results*

   Excellent organizations comprehensively measure and achieve outstanding results with respect to society.

8. *Key results*

   Excellent organizations comprehensively measure and achieve outstanding results with respect to the key elements of their policy and strategy.

Source: EFQM Excellence Model © EFQM, Brussels, Belgium.

THE AGILE MANIFESTO

In February 2001, 17 software developers met to discuss lightweight software development methods. They published the *Manifesto for Agile Software Development* to define the approach now known as agile software development. The Agile Manifesto reads, in its entirety, as follows:

'We are uncovering better ways of developing software by doing it and helping others do it. Through this work we have come to value:

- individuals and interactions over processes and tools
- working software over comprehensive documentation
- customer collaboration over contract negotiation
- responding to change over following a plan.

That is, while there is value in the items on the right, we value the items on the left more.

The meanings of the Manifesto items on the left within the agile software development context are described below.

- Individuals and Interactions – in agile development, self-organization and motivation are important, as are interactions like co-location and pair programming.
- Working software – working software will be more useful and welcome than just presenting documents to clients in meetings.
- Customer collaboration – requirements cannot be fully collected at the beginning of the software development cycle, therefore continuous customer or stakeholder involvement is very important.
- Responding to change – agile development is focused on quick responses to change and continuous development.

Twelve principles underlie the Agile Manifesto, including:

1. Our highest priority is to satisfy the customer through early and continuous delivery of valuable software.
2. Welcome changing requirements, even late in development. Agile processes harness change for the customer's competitive advantage.
3. Deliver working software frequently from a couple of weeks to a couple of months, with a preference to the shorter timescale.
4. Business people and developers must work together daily throughout the project.
5. Build projects around motivated individuals. Give them the environment and support they need, and trust them to get the job done.
6. The most efficient and effective method of conveying information to and within a development team is face-to-face conversation.
7. Working software is the primary measure of progress.
8. Agile processes promote sustainable development. The sponsors, developers and users should be able to maintain a constant pace indefinitely.
9. Continuous attention to technical excellence and good design enhances agility.
10. Simplicity – the art of maximizing the amount of work not done – is essential.
11. The best architectures, requirements, and designs emerge from self-organizing teams.
12. At regular intervals, the team reflects on how to become more effective, then tunes and adjusts its behaviour accordingly.'

Source: Kent Beck, Mike Beedle, Arie van Bennekum, Alistair Cockburn, Ward Cunningham, Martin Fowler, James Grenning, Jim Highsmith, Andrew Hunt, Ron Jeffries, Jon Kern, Brian Marick, Robert C. Martin, Steve Mellor, Ken Schwaber, Jeff Sutherland, Dave Thomas, 2001. *Principles Behind the Agile Manifesto*. Agile Alliance. Retrieved on 6 June, 2010 from http://www.agilemanifesto.org/principles.html.

FURTHER READING

SUSTAINABILITY

Global Reporting Initiative: For more information see: https://www.global reporting.org/reporting/guidelines-online/G3Online/Pages/default.aspx/. Accessed 21 April 2012.

ISO 26000: For more information see: http://www.iso.org/iso/iso_catalogue/ management_and_leadership_standards/social_responsibility/sr_discovering_ iso26000.htm.

UN Global Compact: An overview of the UN Global Compact's ten principles is available at: http://www.unglobalcompact.org/aboutthegc/thetenprinciples/ index.html. Accessed 21 April 2012.

The Natural Step Framework: For more information see: http://www.naturalstep. org.

PROJECTS AND PROJECT MANAGEMENT

Turner, R., Huemann, M., Anbari, F. and Bredillet, C. (2010), *Perspectives on Projects*, Routledge, Abingdon.

SUSTAINABILITY IN PROJECTS AND PROJECT MANAGEMENT

Knoepfel, H. (Ed.) (2010), *Survival and Sustainability as Challenges for Projects*, International Project Management Association, Zurich.

ETHICS IN PROJECTS AND PROJECT MANAGEMENT

Knoepfel, H., Scheifele, D., Staeuble, M. and Witschi, U. (2008), *Values and Ethics in Project Management*, International Project Management Association, Zurich.

Jonasson, H.I. and Ingason, H.T. (2011), *Project Ethics*, Gower, Farnham.

POSITIVE PSYCHOLOGY

Lewis, S. (2001), *Positive Psychology at Work: How Positive Leadership and Appreciative Inquiry Create Inspiring Organizations*, Wiley-Backwell., Chichester.

Lyubomirsky, S. (2007), *The How of Happiness: A Practical Guide to Getting the Life You Want*, Piatkus, London.

INDEX

ADVANCES IN PROJECT MANAGEMENT

Advances in Project Management provides short, state of play guides to the main aspects of the new emerging applications, including: maturity models, agile projects, extreme projects, Six Sigma and projects, human factors and leadership in projects, project governance, value management, virtual teams and project benefits.

CURRENTLY PUBLISHED TITLES

Managing Project Uncertainty, David Cleden 978-0-566-08840-7

Strategic Project Risk Appraisal and Management, Elaine Harris 978-0-566-08848-3

Project-Oriented Leadership, Ralf Müller and J. Rodney Turner 978-0-566-08923-7

Tame, Messy and Wicked Risk Leadership, David Hancock 978-0-566-09242-8

Managing Project Supply Chains, Ron Basu 978-1-4094-2515-1

REVIEWS OF THE SERIES

Managing Project Uncertainty, David Cleden

> *This is a must-read book for anyone involved in project management. The author's carefully crafted work meets all my "4Cs" review criteria. The book is clear, cogent, concise and complete ... it is a brave author who essays to write about managing project uncertainty in a text extending to only 117 pages (soft-cover version). In my opinion, David Cleden succeeds brilliantly. ... For project managers this book, far from being a short-lived stress anodyne, will provide a confidence-boosting tonic. Project uncertainty? Bring it on, I say!*
> *International Journal of Managing Projects in Business*

> *Uncertainty is an inevitable aspect of most projects, but even the most proficient project manager struggles to successfully contain it. Many projects*

overrun and consume more funds than were originally budgeted, often leading to unplanned expense and outright programme failure. David examines how uncertainty occurs and provides management strategies that the user can put to immediate use on their own project work. He also provides a series of pre-emptive uncertainty and risk avoidance strategies that should be the cornerstone of any planning exercise for all personnel involved in project work.

I have been delivering both large and small projects and programmes in the public and private sector since 1989. I wish this book had been available when I began my career in project work. I strongly commend this book to all project professionals.

Lee Hendricks, Sales & Marketing Director,
SunGard Public Sector

The book under review is an excellent presentation of a comprehensive set of explorations about uncertainty (its recognition) in the context of projects. It does a good job of all along reinforcing the difference between risk (known unknowns) management and managing uncertainty (unknown unknowns – "bolt from the blue"). The author lucidly presents a variety of frameworks/ models so that the reader easily grasps the varied forms in which uncertainty presents itself in the context of projects.

VISION: The Journal of Business Perspective (India)

Cleden will leave you with a sound understanding about the traits, tendencies, timing and tenacity of uncertainty in projects. He is also adept at identifying certain methods that try to contain the uncertainty, and why some prove more successful than others. Those who expect risk management to be the be-all, end-all for uncertainty solutions will be in for a rude awakening.

Brad Egeland, Project Management Tips

Strategic Project Risk Appraisal and Management, Elaine Harris

Elaine Harris's volume is timely. In a world of books by "instant experts" it's pleasing to read something by someone who clearly knows their onions, and has a passion for the subject. In summary, this is a thorough and engaging book.

Chris Morgan, Head of Business Assurance for Select Plant Hire,
Quality World

As soon as I met Elaine I realised that we both shared a passion to better understand the inherent risk in any project, be that capital investment, expansion capital or expansion of assets. What is seldom analysed are the components of knowledge necessary to make a good judgement, the impact of our own prejudices in relation to projects or for that matter the cultural elements within an organisation which impact upon the decision making process. Elaine created a system to break this down and give reasons and logic to both the process and the human interaction necessary to improve the chances of success. Adopting her recommendations will improve teamwork and outcomes for your company.

Edward Roderick Hon. LLD, former CEO Christian Salvesen plc

Tame, Messy and Wicked Risk Leadership, David Hancock

This book takes project risk management firmly onto a higher and wider plane. We thought we knew what project risk management was and what it could do. David Hancock shows us a great deal more of both. David Hancock has probably read more about risk management than almost anybody else; he has almost certainly thought about it as much as anybody else and he has quite certainly learnt from doing it on very difficult projects as much as anybody else. His book draws fully on all three components. For a book which tackles a complex subject with breadth, insight and novelty – it's remarkable that it is also a really good read. I could go on!

Dr Martin Barnes CBE FREng, President,
The Association for Project Management

This compact and thought-provoking description of risk management will be useful to anybody with responsibilities for projects, programmes or businesses. It hits the nail on the head in so many ways, for example by pointing out that risk management can easily drift into a checklist mindset, driven by the production of registers of numerous occurrences characterised by the Risk = Probablity × Consequence equation. David Hancock points out that real life is much more complicated, with the heart of the problem lying in people, so that real life resembles poker rather than roulette. He also points out that while the important thing is to solve the right problem, many real-life issues cannot be readily described in a definitive statement of the problem. There are often interrelated individual problems with surrounding social issues and he describes these real-life situations as "Wicked Messes". Unusual terminology, but definitely worth the read, as much for the overall problem description as for the recommended strategies for getting to grips with real-life risk management. I have no hesitation in recommending this book.

Sir Robert Walmsley KCB FREng, Chairman of the Board of the Major
Projects Association

In highlighting the complexity of many of today's problems and defining them as tame, messy or wicked, David Hancock brings a new perspective to the risk issues that we currently face. He challenges risk managers, and particularly those involved in project risk management, to take a much broader approach to the assessment of risk and consider the social, political and behavioural dimensions of each problem, as well as the scientific and engineering aspects with which they are most comfortable. In this way, risks will be viewed more holistically and managed more effectively than at present.

Dr Lynn T. Drennan, Chief Executive Alarm,
The Public Risk Management Association

ABOUT THE EDITOR

Professor Darren Dalcher is founder and Director of the National Centre for Project Management, a Professor of Software Project Management at Middlesex University and Visiting Professor of Computer Science at the University of Iceland. Professor Dalcher has been named by the Association for Project Management as one of the top 10 'movers and shapers' in project management. He has also been voted *Project Magazine*'s Academic of the Year for his contribution in 'integrating and weaving academic work with practice'.

Professor Dalcher is active in numerous international committees, steering groups and editorial boards. He is heavily involved in organising international conferences, and has delivered many keynote addresses and tutorials. He has written over 150 papers and book chapters on project management and software engineering. He is Editor-in-Chief of *Software Process Improvement and Practice*, an international journal focusing on capability, maturity, growth and improvement.

Professor Dalcher is a Fellow of the Association for Project Management and the British Computer Society, and a Member of the Project Management Institute, the Academy of Management, the Institute for Electrical and Electronics Engineers and the Association for Computing Machinery. He is a Chartered IT Practitioner. He is a member of the PMI Advisory Board responsible for the prestigious David I. Cleland project management award, and of the APM Professional Development Board.

National Centre for Project Management
Middlesex University
College House
Trent Park
Bramley Road
London N14 4YZ
Email: ncpm@mdx.ac.uk
Phone: +44 (0)20 8411 2299